Mathematical Modeling of Inland Vessel Maneuverability Considering Rudder Hydrodynamics

Jialun Liu

Mathematical Modeling of Inland Vessel Maneuverability Considering Rudder Hydrodynamics

 Springer

Jialun Liu
Wuhan University of Technology
Wuhan, Hubei, China

ISBN 978-3-030-47477-5   ISBN 978-3-030-47475-1   (eBook)
https://doi.org/10.1007/978-3-030-47475-1

This Springer imprint is published by the registered company Springer Nature Switzerland AG
The registered company address is: Gewerbestrasse 11, 6330 Cham, Switzerland

To my wife Shijie,
my muse.
Jialun Liu

Preface

The motivation for writing this monograph was to provide knowledge about inland vessel maneuverability and ship rudders that affect ship performance to a large extent. The ideas are put forward together by a broad literature review of the-state-of-the-art research in this field and the specified work on the impacts of rudders on inland vessel maneuvering performance. The contents were partly published in my doctoral thesis entitled 'Impacts of Rudder Configurations on Inland Vessel Maneuverability' from Delft University of Technology [1]. The manuscript was rephrased and updated to provide more detailed information than the original thesis. This monograph is intended as a reference monograph for modeling inland vessel motions and selecting proper rudder configurations to a wide range of researchers, engineers, and students.

Ship maneuverability is fundamental for the navigation safety of ships. Furthermore, through the equipment used for maneuvering, it also affects the investment, operation, and maintenance cost of these ships. Ships are primarily designed from an economic point of view. To ensure and improve maritime efficiency, research on inland vessel maneuverability deserves more attention than the present situation. Most of the research on maneuverability has been performed for seagoing ships. Since sailing conditions and ship particulars between seagoing ships and inland vessels are different, the impacts of these differences on maneuvering prediction and evaluation should be carefully considered, which are addressed in this monograph.

Inland vessels should be designed in such a way that they should always be capable of maneuvering without significantly harming the cost-effectiveness of operations. One of the biggest differences between seagoing ships and inland vessels is the rudder configuration. Conventionally, seagoing ships have similar single-rudder configurations while inland vessels have more complex multiple-rudder configurations. Although multiple-rudder configurations can have a positive effect on maneuverability, they often have a negative effect on resistance and, therefore, also a negative effect on fuel consumption.

Quantitative impacts of the rudder configuration on ship maneuverability have not been fully understood, especially for multiple-rudder configurations with complex rudder profiles. These differences in the rudder configuration may significantly change the ship maneuvering behaviors and, therefore, should require further research, especially for inland vessels that commonly equip with more than two rudders. Moreover, to compare and evaluate the maneuvering performance of inland vessels with different configurations, the existing maneuvering tests and standards for inland vessels are less elaborate than those for seagoing ships.

An accurate estimation of rudder forces and moments is needed to quantify the impacts of the rudder configurations on ship maneuvering performance. This monograph explained the fundamentals of Computational Fluid Dynamics and applied best practices of the Reynolds-Averaged Navier-Stokes (RANS) method to obtain rudder hydrodynamic characteristics and integrated the RANS results into maneuvering models. Additionally, new maneuvers and criteria have been proposed for the prediction and evaluation of inland vessel maneuverability. Simulations of ships with various rudder configurations were conducted to analyze the impacts of rudder configurations on ship maneuverability in different classic and proposed test maneuvers. Accordingly, guidance on rudders for inland vessel maneuverability has been summarized for practical engineers to make proper design choices.

Through the research presented in this monograph, it is clear that different rudder configurations have different hydrodynamic characteristics, which are influenced by the profile, the parameters, and the type of a specific configuration. New regression formulas have been proposed for naval architects to quickly estimate the rudder induced forces and moments in maneuvering. Furthermore, an integrated maneuvering model has been proposed and validated for both seagoing ships and inland vessels. Using the proposed regression formulas and maneuvering model, the impacts of rudder configurations on inland vessel maneuverability have been studied.

The maneuvering performance of a typical inland vessel can be improved by 5 to 30% by changing the rudder configuration. The rudder configuration should be capable of providing sufficient maneuvering forces and then optimized to reduce the rudder induced resistance. In general, well-streamlined profiles are good for efficiency but not as good as high-lift profiles for effectiveness. As a summary, the ship maneuvering performance can be improved by using effective profiles, enlarging the total rudder area, accelerating the rudder inflow velocity, increasing the effective rudder aspect ratios, and enlarging the spacing among multiple rudders.

I would like to acknowledge my doctoral supervisors, Dr. Ir. Robert Hekkenberg and Prof. Hans Hopman from Delft University of Technology, for their support and contribution to this monograph. Special thanks go to Mr. Frans Quadvlieg from Maritime Research Institute Netherlands (MARIN) for advising and improving the presented research. I am grateful to the members of my doctoral committee, Prof. Dr. Hironori Yasukawa (Hiroshima University), Prof. Dr.-Ing Bettar Ould el

Moctar (University of Duisburg-Essen), Prof. Dr. Ir. Rene Huijsmans (Delft University of Technology), Prof. Dr. Dirk Roekaerts (Delft University of Technology), and Dr. Henk de Koning Gans (Delft University of Technology), for reviewing and commenting on the thesis that lays a good foundation for this monograph.

I would also like to say thanks to Prof. Xinping Yan, Prof. Xiumin Chu, Prof. Xuming Wang, and Dr. Feng Ma that help me to continue the research topic and further extend it for autonomous ships in Wuhan University of Technology. Contributions from Mr. Bingqian Zhao are acknowledged for providing data of inland vessels in China for explanations and validation. Furthermore, the effort of students, including, but not limited to, Song Zhang, Suli Lu, Jinyu Kan, and Yanyun Zhang, who helped to review and edit the content is acknowledged.

The contents in the monograph were initially supported by China Scholarship Council (201206950025) from September 2012 to February 2017. The research is further supported by National Key R&D Program of China (2018YFB1601505), Research on Intelligent Ship Testing and Verification ([2018]473), National Natural Science Foundation of China (51709217), Natural Science Foundation of Hubei Province (2018CFB640), and State Key Laboratory of Ocean Engineering (Shanghai Jiao Tong University) (1707) since March 2017.

Wuhan, China   Jialun Liu
March 2020

Reference

1. Liu, J. (2017). Impacts of Rudder Configurations on Inland Vessel Manoeuvrability. PhD thesis, Delft University of Technology

Contents

About the Author

Dr. Jialun Liu is an Associate Professor in Intelligent Transportation Systems Research Center, Wuhan University of Technology and a member of National Engineering Research Center for Water Transport Safety. From 2006 to 2010, he studied at Wuhan University of Technology, Wuhan, China, specializing in Navigation Technology. From 2008 to 2010, he minored in Japanese at Huazhong University of Science and Technology, Wuhan, China. He obtained his BSc and BA degrees in 2010 from these two universities, respectively. After that, he was recommended to continue master education at Wuhan University of Technology in Traffic Engineering and Control. In 2012, he was founded by China Scholarship Council and started his Ph.D. research at Delft University of Technology, Delft, the Netherlands. In 2013, he received his MSc degree. From 2012 to 2017, Jialun Liu worked at the section of Ship Design, Operation, and Production in the Department of Marine and Transportation Technology. From March 2017, he started working at Wuhan University of Technology. He received "Chutian Excellent Young Scholar" in 2017 and got promoted to be an associate professor. He has been appointed as a reviewer of Ocean Engineering, Ships and Offshore Structures, Journal of Marine Science and Technology, International Journal of Naval Architecture and Ocean Engineering, and other international journals and conferences. He has interests in maneuvering modeling and motion control of ships, applications of computational fluid dynamics methods to maritime engineering, developments of autonomous ships, and functional testing of smart ships.

March 2020

Glossary

Abbreviations

ABS	American Bureau of Shipping
BV	Bureau Veritas
CCNR	Central Commission for the Navigation of the Rhine
CCS	China Classification Society
CFD	Computational Fluid Dynamics
COG	Centre of Gravity
DNS	Direct Numerical Simulation
DOF	Degree of freedom
EC	European Commission
EEDI	Energy Efficiency Design Index
EU	European Union
HSVA	Hamburgische Schiffbau Versuchsanstalt GmbH
IFS	Institute für Schiffbau
IMO	International Maritime Organization
IWT	Inland Waterway Transport
LES	Large Eddy Simulation
MARIN	Maritime Research Institute Netherlands
OSA	Overshoot angle
PMM	Planar Motion Mechanism
QPI	Inner rudder of the port side twin-rudder unit
QPO	Outer rudder of the port side twin-rudder unit
QSI	Inner rudder of the starboard side twin-rudder unit
QSO	Outer rudder of the starboard side twin-rudder unit
RANS	Reynolds-Averaged Navier-Stokes
RPS	Revolutions per second
SPSR	Single-propeller single-rudder
SPTR	Single-propeller twin-rudder
TPQR	Twin-propeller quadruple-rudder

TPTR   Twin-propeller twin-rudder
TSS   Traffic Separation Scheme

Greek Symbols

α Angle of attack, (rad)
α_0 Incidence for zero lift, (rad)
α_R Effective rudder angle, (rad)
$\bar{\delta}$ Average applied rudder angle, (rad)
β Ship drift angle on midship (rad)
β_P Drift angle at propeller position, (rad)
β_R Drift angle at rudder position, (rad)
δ Applied rudder angle, (rad)
Δ Relative difference, (%)
δ_h Hydrodynamic inflow angle of the rudder, (rad)
$\dot{\delta}$ Rudder turning rate, (rad s^{-1})
ε Dissipation of the turbulent energy, (–)
ε_R Ratio of wake fraction of propeller to wake fraction of rudder, (–)
η Ratio of rudder consumed power to the total power consumed by the hull
               and the rudder, (–)
η_R Ratio of propeller diameter to rudder span, $\eta_R = D_P/B_R$, (–)
k_P Impact factor of the propeller slipstream on the rudder hydrodynamics, (–)
k_R Impact factor of the endplates on the rudder hydrodynamics, (–)
γ_R Flow straightening coefficient of the rudder, (–)
κ_R Experimental constant for expressing u_R, (–)
λ Model scale, (–)
Λ_E Rudder effective aspect ratio, (–)
Λ_G Rudder geometric aspect ratio, (–)
μ Dynamic viscosity of water, (N s m^{-2})
∇ Ship displacement volume, (m^3)
ω Specific rate of dissipation of turbulence kinetic energy, (–)
ψ Heading angle, (rad)
ψ_{O1} First overshoot angle, (rad)
ψ_{O2} Second overshoot angle, (rad)
ρ Density of water, (kg m^{-3})
σ Root mean squared error (standard error) of the regression coefficients, (–)
σ_T Average absolute deviation of the turning criteria, (–)
σ_Z Average absolute deviation of the zigzag criteria, (–)

Roman Symbols

A_D	Advance in the turning maneuver, $A_D' = A_D/L$, (m)
a_H	Rudder force increase factor, (–)
A_m	Ship midship section area, (m^2)
A_R	Rudder lateral area without the horn part, (m^2)
A_{RP}	Rudder lateral area in the propeller slipstream, (m^2)
T_D	Tactical diameter of turning circle test, $T_D' = T_D/L$, (m)
B	Ship width at water level, (m)
B_R	Rudder span, (m)
C_b	Block coefficient, (–)
C_D	Drag coefficient, (–)
C_L	Lift coefficient, (–)
C_L/C_D	Lift to drag ratio, (–)
C_m	Midship coefficient, (–)
C_N	Normal force coefficient, (–)
C_p	Prismatic coefficient, (–)
C_R	Rudder chord length, (m)
C_T	Axial force coefficient, (–)
C_{D_0}	Drag coefficient at zero angle of attack, (–)
C_{L_0}	Lift coefficient at zero angle of attack, (–)
$C_{YR\alpha}$	Gradient of the rudder induced side force coefficient, (–)
C_{YR}	Rudder induced side force coefficient, (–)
D_P	Propeller diameter, (m)
d_R	Clearance between rudder tip and bottom of waterway, $d_R' = d_R/B_R$, (m)
F_D	Rudder drag force, (N)
F_L	Rudder lift force, (N)
F_N	Rudder normal force, (N)
F_R	Rudder resultant force, (N)
F_T	Rudder tangential force, (N)
F_X	Longitudinal component of rudder induced forces, (N)
F_Y	Lateral component of rudder induced forces, (N)
Fr	Froude number, (–)
H	Water depth, (m)
I_z	Moment of inertial, (kg m^2)
J_P	Propeller advance ratio, $J_P = V_P/nD_P$, (–)
J_z	Added moment of inertial, (kg m^2)
k	Turbulent kinetic energy, (–)
k^{C_D}	Regression coefficients of twin-rudder drag coefficients, (–)
k^{C_L}	Regression coefficients of twin-rudder lift coefficients, (–)
k_Λ	Impact factor of the rudder aspect ratio on the rudder hydrodynamics, (–)
K_T	Propeller thrust coefficient, (–)
L	Ship length between perpendiculars, (m)
m	Ship's mass, (kg)

m_x	Ship's added mass in x-direction, (kg)
m_y	Ship's added mass in y-direction, (kg)
N	Total hydrodynamic moment acting on the ship, (N m)
n	Propeller revolution rate, (s^{-1})
N_H	Hydrodynamic moment due to hull acting on the ship, (N m)
N_P	Hydrodynamic moment due to propeller acting on the ship, (N m)
n_P	Number of propellers, (–)
N_R	Hydrodynamic moment due to rudder acting on the ship, (N m)
n_R	Number of rudders, (–)
n_T	Number of performed turning maneuvers, (–)
n_Z	Number of performed zigzag maneuvers, (–)
P	Power consumed by the rudder induced resistance, (kW)
$P(\delta)$	Probability distribution of each applied rudder angle, (–)
P_P	Propeller pitch, (m)
$\dot{r}$	Yaw acceleration around midship, (rad s^{-2})
r	Yaw rate around midship, (rad s^{-1})
R^2	R-squared (coefficient of determination) of the regression coefficients, (–)
r_C	Yaw rate in steady turn, (rad s^{-1})
Re	Reynolds number, (–)
S	Wetted surface, (m)
S_H	Wetted surface of the ship hull, (m)
T	Ship draft, (m)
t	Rudder profile section thickness, (m)
T_P	Propeller thrust, (N)
t_P	Propeller thrust deduction, (–)
t_R	Steering resistance deduction factor, (–)
t_{O1}	Time to the first overshoot angle, (s)
t_{O2}	Time to the second overshoot angle, (s)
$\dot{u}$	Ship acceleration in x-direction, (m s^{-2})
u	Forward speed in x-direction, $u = V\cos\beta$, (m s^{-1})
u_R	Longitudinal velocity of the inflow to rudder, (m s^{-1})
$\dot{v}$	Ship acceleration in y-direction, (m s^{-2})
V	Ship velocity on midship, $V = \sqrt{u^2 + v^2}$, (m s^{-1})
v	Lateral speed in y-direction on midship, $v = -V\sin\beta$, (m s^{-1})
V_A	Propeller advance speed, (m s^{-1})
V_C	Speed in steady turn, (m s^{-1})
V_P	Propeller advance speed, (m s^{-1})
V_R	Rudder inflow velocity, (m s^{-1})
v_R	Lateral velocity of the inflow to rudder, (m s^{-1})
V_S	Service speed, (m s^{-1})
w_P	Wake factor at propeller position in maneuvering, (–)
w_R	Wake factor at rudder position in maneuvering, (–)
w_{P_0}	Wake factor at propeller position in straight moving, (–)
X	Total hydrodynamic force acting on midship in x-direction, (N)

x_G Longitudinal position of center of gravity in$o$-xyz, (m)
X_H Hydrodynamic force due to hull acting on midship in x-direction, (N)
x_H Longitudinal position of acting point of additional lateral force, (m)
X_P Hydrodynamic force due to propeller acting on midship in y-direction, (N)
x_P Longitudinal position of propeller in$o$-xyz, (m)
X_R Hydrodynamic force due to rudder acting on midship in x-direction, (N)
x_R Longitudinal position of rudder in$o$-xyz, (m)
x_{PR} Longitudinal propeller–rudder separation, (m)
Y Total hydrodynamic force acting on midship in y-direction, (N)
y^+ Non-dimensional wall distance, (–)
y_i Distance from the wall to the first mesh point in mesh generation, (–)
Y_R Hydrodynamic force due to rudder on midship in y-direction, (N)
y_R Relative position of the rudder to the propeller shaft, (m)
y_{PR} Lateral propeller–rudder separation, (m)
y_{TR} Spacing between the twin rudders, (m)
y_{TU} Spacing between the two twin-rudder units, (m)
z_{PR} Vertical propeller–rudder separation, (m)

Subscripts

AH   Route from Antwerp, Belgium to Vlaardingen, the Netherlands
H   Hull
P   Propeller
R   Rudder
VH   Route from Vlaardingen, the Netherlands to Hamm, Germany

Superscripts

Exp   Experiment
P   Port side
S   Starboard Side
Sim   Simulation
TP   Port side of a twin-ruder configuration
TS   Starboard side of a twin-ruder configuration

Chapter 1
Introduction

Inland shipping plays an important role in transport. It is reliable, low-cost, and environment-friendly. Nowadays, inland navigation plays an increasingly important role in logistics [1]. To maintain and further develop this important role, safe and efficient inland vessels are prerequisites. Inland vessels should always be capable of safe maneuvering without significantly harming the cost-effectiveness of operations. Options to influence maneuverability through the design of inland vessels, such as the hull form and the main dimensions, are commonly constrained due to the limits of inland waterways. One of the possibilities that are left to improve ship maneuverability is to optimize rudder configurations. Additionally, practical engineers and authority officers necessitate test maneuvers and related criteria to predict and evaluate ship maneuverability for design and management. This chapter presents the background and significance of this monograph.

1.1 Applied Terminologies

This chapter provides a background on the importance of inland navigation. When considering terms like design and performance, many definitions appear in literature and general use. To avoid ambiguity, in this monograph, the following terminologies have been chosen:

- An **inland vessel** is a self-propelled motor ship that sails in inland waterways, such as rivers, canals, and lakes.
- **Ship maneuverability** is the ability of a ship to react to certain navigation commands, which include, but not limited to, turning, evading, yaw checking, course keeping, and stopping.

© Springer Nature Switzerland AG 2020
J. Liu, *Mathematical Modeling of Inland Vessel Maneuverability Considering Rudder Hydrodynamics*, https://doi.org/10.1007/978-3-030-47475-1_1

- A **test maneuver** is the scenario that describes the conditions to start and end a specific maneuvering test. Different test maneuvers emphasize different aspects of ship maneuverability.
- A **rudder configuration** is the combination of rudder design choices related to the rudder profile, the rudder parameters, the rudder type, the number of rudders, the rudder appendages (endplates), and the relative positions to the hull, the propeller, and other rudders.
- A **rudder profile** is the sectional shape of a rudder in two dimensions.
- A **rudder parameter** is one of the rudder design choice, such as area, thickness, span, chord, and aspect ratios.
- A **rudder type** depends on the structure, the position of the rudder stock (unbalanced, semi-balanced, or balanced), and the structural rudder-hull connection (the number of pintles, no skeg, semi-skeg, or full-skeg).
- **Rudder effectiveness** is judged by the amount of rudder induced maneuvering force that generates the yaw moment. For open water rudder hydrodynamics, the rudder effectiveness is evaluated by the normal force coefficient.
- **Rudder efficiency** is defined as the ratio of the rudder induced useful force (lift in open water or lateral force in maneuvering) to useless force (drag in open water or longitudinal force in maneuvering). In this monograph, the lift to drag ratio is taken as an indicator of the rudder efficiency for open water tests.

1.2 Inland Waterway Transport

Inland waterway transport (IWT) plays a significant role in the modal split of transport for a long time and still does [2, Sect. 1.1]. It takes a large modal split in countries, including, but not limited to, Brazil (the Paraguay-Paraná inland waterway), Bulgaria and Romania (the Danube), China (the Yangtze River, the Pearl River, the Xijiang River), Germany and the Netherlands (the Rhine and the Danube), and United States of America (the Mississippi) [3–5]. The Netherlands, Germany, and Belgium are European countries with the largest capacity of inland waterway transport. Over half of the international country flows for transport are loaded and unloaded among these three countries [6]. In Germany, up to 240 million tonnes of bulk goods are transported per year via inland waterways, which almost equals 75% of the goods transported by railway [7]. Based on the data from 2005 to 2013, the IWT of Belgium increased from 14.5 to 20.5%, which was larger than the modal split share of railway [4]. Additionally, the annual volume of IWT in the United States is about 600 million tonnes [8].

China has over 110000 km navigable inland waterways as shown in Fig. 1.1. The main channels are the Yangtze River, the Pearl River, and the Xijiang River. These waterways are wide, stable, and ice-free all year round, thus inland waterway transport is flourishing [9–11]. Transport of containers, liquid bulk cargo, and automobiles are expected to increase most rapidly [9]. Cargoes transported in the mainstream and

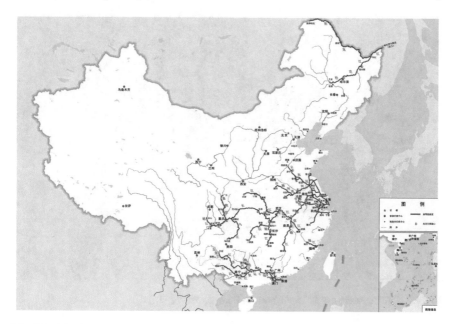

Fig. 1.1 Inland waterways and main ports in China. *Source* [3]

tributary of the Yangtze River has been first in the world. The total cargo increased from 690 million t (155 billion tkm) in 2000 to 1160 million t (303 billion tkm) in 2006, and it is expected to reach 2350 million t (658 billion tkm) in 2020 [3]. The main challenges for the future are modernization and standardization of fleets, continuous channel upgrades, integration and commercialization of family-operated vessels to shipping companies, and awareness of sustainable development [9, 11].

Europe has a total navigable waterway of about 51700 km, including canals, rivers, and lakes, of which approximately 20000 km are mainly concentrated in the Netherlands, Germany, Belgium, France, and Austria [12]. Almost all important industrial regions in Western Europe are reachable through inland waterways as shown in Fig. 1.2. European Inland Waterways are classified as CEMT Class I–VII [13]. For different types of cargo and waterway capacities, ship dimensions vary from the smallest ship of 30 m (350 t) to the largest ship of 135 m (5000 t) [13]. On the other hand, the practical upper limits on length, beam, and draft for motor vessels in the Rhine may be set at 186.5, 22.9, and 4.5 m [2], respectively. Thus, there is the possibility to enlarge inland vessels. Accordingly, enlargement of inland vessels puts more challenges on inland ship maneuverability.

Up to 2017, there are about 119300 ships (including inland vessels and seagoing ships) registered in the 14 provinces along the Yangtze River. The net tonnage for cargo transport is 1.97 million tonnes and the rated passenger capacity is 566200. The capacity of container transport is 1591500 TEU. The total propulsion power of the fleet is 50997900 kW. Container transport shows an increase of 11% while cargo and

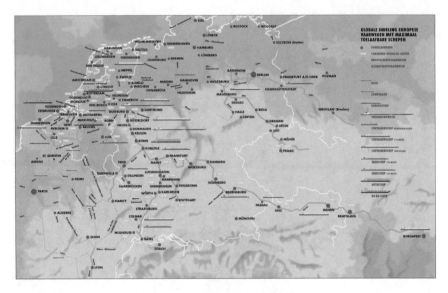

Fig. 1.2 Inland waterways and main ports in Europe. *Source* [12]

passenger transport drops 1 and 1% respectively compared to the previous five years. The future trend of inland vessels developments is standardization, specialization, and enlargement.

Up to now, inland shipping capacity in European Union (EU) has developed to 545 million tonnes per year in 2014, with a fleet counted around 13000 vessels [4]. Figure 1.3 shows that the modal split of inland navigation is smaller than road transport but comparable with rail transport. The European Union (EU) sets an objective of shifting traffic from roads to rail and water-borne transport for cost savings, reducing pollution, and increasing transport safety [14]. However, according to [15], the EU inland waterway transport strategies have not been effectively implemented and policy objective has not been achieved yet. Therefore, projects that provide benefits for inland waterway transport are highly encouraged [15]. To further develop inland navigation, on the one hand, infrastructure bottlenecks in waterways should be eliminated, on the other hand, inland vessels themselves should be capable of sailing safely and efficiently. Safety and efficiency of inland vessels justify significant research attention.

In recent years, the inland navigation split of inland transport also increases significantly in Bulgaria, Luxembourg, and Romania as illustrated in Fig. 1.4. It is noticeable that the modal split of the inland navigation in the Netherlands accounts for around 30% of the total transported cargo and the percentage is still increasing. As a representative, the Netherlands has an inland shipping fleet of around 7000 vessels, which is the largest and most modern one in Europe, taking around 35% market share in total Dutch freight (Fig. 1.4). The inland shipping sector is indispensable for the Netherlands, taking a modal split share of more than 35% of the Dutch freight [4].

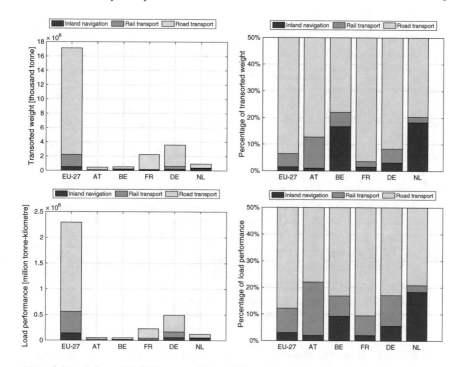

Note of abbreviations: EU-27 (European Union of 27 member states), AT (Austria), BE (Belgium), FR (France), DE (Germany), and NL (The Netherlands).

Fig. 1.3 Modal split of transported weight and load performance in 2011. *Source* [4]

Container shipping, which has an increasing number of 4.5 million TEUs, accounts for 35% of the whole freight hauled in inland waterways. Traditionally, an estimate of around 75% ships is owned and maintained by the family business.

1.3 Inland Vessel Maneuverability

Inland navigation is the transport with vessels along inland waterways. The waters of inland and sea differ in channel width, water depth, and air clearance leading to the diversity of inland vessels and seagoing ships in hull forms, ratios of the main particulars, and facilitated equipment. These differences of ships and environment will be further addressed in Sect. 5.1. Reference types of inland vessels in the Yangtze River and the River Rhine are shown in Fig. 1.5. Inland vessels commonly have a more blunt hull and sail at a slower speed than seagoing ships, which may lead to different maneuvering performance and require attention with the equipment. However, less attention has been paid to inland vessel design than seagoing ships due to a lack of investment and profits.

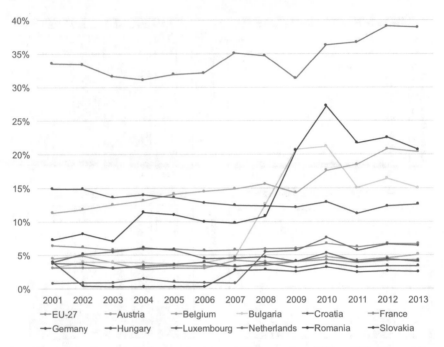

Fig. 1.4 Modal split trend of inland navigation per EU country from 2001 to 2013. *Source* [4]

Developments and innovations enable cost reduction, improvement of environmental performance, access niche markets, and compliance with new regulations
[5]. The safety of the ship is a prerequisite for efficient and reliable inland waterway
transport. Nevertheless, ships have been primarily designed from an economic point
of view [16]. Due to the potentially high cost of improving maneuverability, a wise,
well-balanced, compromise between performance and interest has to be made [17].
Furthermore, the reasons why the maneuverability of inland vessels deserves special
attention are as follows:

- Inland waterways are often narrow, bendy, and shallow. Artificial structures, such
  as bridges, wharves, and locks, also put limits on the navigational area. Besides,
  strong currents may happen in conjunctions of channels, especially in flood sea-
  sons. Thus, ship maneuvering performance is crucial to ensure safe navigation.
- Inland waterways are frequently busy, making the navigation situation complex.
  A lot of encountering and overtaking may happen in day-to-day operations which
  require inland vessels to be well maneuverable. More specifically, inland vessels
  should be able to quickly and effectively respond to the orders of the skippers.
- Unlike seagoing ships that may get assistance from tugs in hard maneuvering
  situations, inland vessels sail independently most of the time. Additional consideration of low-speed maneuverability should be taken into account. Furthermore,
  the enhancement of low-speed maneuverability should not harm cruising-speed
  performance.

(a) Inland cargo vessel (the Yangtze).   (b) Inland chemical tanker (the Yangtze).

(c) Inland container vessel (the Yangtze).   (d) Inland cruise vessel (the Yangtze).

(e) Inland Ro-Ro vessel (the Yangtze).   (f) Inland ferry (the Yangtze).

(g) Inland cargo vessel (the Rhine).   (h) Inland chemical tanker (the Rhine).

Fig. 1.5 Inland vessels in the Yangtze River. *Source* Own photographs, Mr. Bingqian Zhao from China Classification Society, and Dr. Robert Hekkenberg from Delft University of Technology

- Inland vessels customarily equip multiple rudders with a large variety of profiles, parameters, types, and relative positions. Impacts of these rudder choices on ship maneuverability have not been sufficiently understood yet. Furthermore, a rudder may increase the total resistance by 1% at the neutral position and 2 to 6% at moderate angles [18]. Correspondingly, optimizing the rudder profile and type can reduce the total resistance by 2 to 8% [19].

There are many ways to improve ship maneuverability, such as increasing the rudder area, upgrading the thrust power, reducing the ship weight, and applying active steering devices. In practice, inland vessels have few choices in main particulars and hull forms because the main particulars are constrained by the width and depth of waterways, canals, and, especially, locks. In Europe, for each class of inland waterways, the dimensions of inland vessels are regulated [13]. Additionally, to maximize the capacity of the ships, hull forms of inland vessels typically have similar large block coefficients (C_b), length to ship width ratios (L/B), and ship width to draft ratios (B/T) [16]. Therefore, one of the remaining possibilities to significantly improve ship maneuverability without changing the main dimensions and hull forms is to optimize rudder configurations.

In general, rudders have significant impacts on ship maneuvering performance as well as an underestimated potential for fuel savings [19]. Molland and Turnock [20, p. 86] concluded the role of rudders in maneuvering as follows: "The rudder has to be able to develop sufficient side force to maintain the ship on a straight course at typical service speeds, to change course at service speed, and to maneuver at slower speeds. In the case of course keeping, interest is centered on minimizing deviations from the set course. In the cases of changes in course or maneuvering, interest centers on the ease and rapidity with which a ship takes up a new course. The effectiveness of the rudder in these two situations will depend on the directional stability of the ship".

1.4 Arrangements of the Subjects

In this monograph, the impacts of rudder configurations on inland vessel maneuverability will be analyzed and guidance on rudder design choices will be proposed. Figure 1.6 describes arrangements of the subjects in this monograph and indicates the read before relations of the chapters. This monograph is organized as follows:

- **This chapter** presents background information and briefly introduces concepts of inland vessel maneuverability and ship rudders to clarify the topics in this monograph.
- **Chapter** 2 presents a state-of-the-art research on ship rudders. The design parameters of a rudder are elaborated and how they may affect the rudder performance.

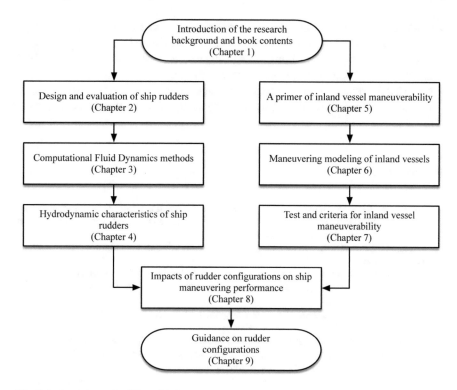

Fig. 1.6 Arrangements of the subjects

- **Chapter** 3 introduces the applied Computational Fluid Dynamics (CFD) methods. Then, these CFD methods are applied to obtain the hydrodynamic characteristics of various rudder configurations in Chap. 4.
- **Chapter** 4 presents the impacts of the rudder profile, the rudder parameters, the number of rudders, and the relative position of rudders on the hydrodynamic coefficients of single, twin, and quadruple rudders. Accordingly, new regression equations are formulated for maneuvering modeling.
- **Chapter** 5 introduces the external and internal factors of a ship to be considered in maneuverability study. Furthermore, a primer introduction is given to understand the mathematical modeling and force analysis.
- **Chapter** 6 presents mathematical maneuvering models and modeling approaches. Using the hydrodynamic characteristics obtained in Chap. 4.
- **Chapter** 7 describes classic and proposed test maneuvers and related criteria for inland vessels. These test maneuvers and criteria are used to evaluate inland vessel maneuverability and compare the maneuvering performance of ships with various rudder configurations.

- **Chapter** 8 studies the impacts of rudder configurations on ship maneuverability. Integrating the results of Chap. 4 into the mathematical models described in Chap. 6, maneuvering simulations of the reference ships are carried out. The simulations are done in test maneuvers that are proposed in Chap. 7.
- **Chapter** 9 provides guidance on rudder design choices. For certain maneuverability and resistance performance, the possible design choices of rudders are listed.

References

1. Bureau Voorlichting Binnenvaart (2014) The power of inland navigation: the future of freight transport and inland navigation in Europe 2013–2014. Technical report, Bureau Voorlichting Binnenvaart, Rotterdam, The Netherlands
2. Hekkenberg R (2013) Inland ships for efficient transport chains. PhD thesis, Delft University of Technology
3. Ministry of transport of the People's Republic of China (2007) Layout planning of China waterways and ports (in Chinese). http://www.gov.cn/gzdt/2007-07/20/content_691664.htm. Accessed 20 Nov 2015
4. Eurostat (2015) Modal split of freight transport. http://ec.europa.eu/eurostat/data/database. Accessed 20 Nov 2015
5. Hekkenberg R, Liu J (2016) Developments in inland waterway vessels. In: Wiegmans B, Konings R (eds) Inland waterway transport: challenges and prospects. Routledge studies in transport analysis, vol 6. Routledge, London, pp 142–167
6. Eurostat (2016) Inland waterway transport statistics. http://ec.europa.eu/eurostat/statistics-explained/index.php/Inland_waterway_transport_statistics. Accessed 30 Sept 2016
7. Federal Ministry of Transport and Digital Infrastructure (2016) Waterways as transport routes. http://www.bmvi.de/SharedDocs/EN/Artikel/WS/waterways-as-transport-routes.html. Accessed 30 Sept 2016
8. World Wide Inland Navigation Network (2015) USA inland waterways. http://www.wwinn.org/us-inland-waterways. Accessed 25 Nov 2015
9. Amos P, Jia D, Ning T, Sun J, Fei W (2009) Sustainable development of inland waterway transport in China. Technical report, The World Bank and The Ministry of Transport, People's Republic of China
10. Kervezee I (2011) Inland waterway transport in the Pearl River Basin. Technical report, Consulate-General of the Kingdom of The Netherlands in Guangzhou, Guangzhou, China
11. World Wide Inland Navigation Network (2015) China inland waterways. http://www.wwinn.org/china-inland-waterways. Accessed 25 Nov 2015
12. Bureau Voorlichting Binnenvaart (2015) Waterways in Europe. http://www.bureauvoorlichtingbinnenvaart.nl/inland-navigation-promotion/basic-knowledge/waterways. Accessed 30 Sept 2015
13. European Conference of Ministers of Transport (1992) Resolution no. 92/2 on new classification of inland waterways
14. European Commission (2011) COM/2011/0144 final White Paper 'Roadmap to a single European transport area – Towards a competitive and resource efficient transport system'
15. European Court of Auditors (2015) Special report no 1/2015 'Inland waterway transport in Europe: no significant improvements in modal share and navigability conditions since 2001'
16. Quadvlieg F (2013) Mathematical models for the prediction of manoeuvres of inland ships: does the ship fit in the river? In: Rigo P, Wolters M (eds) Smart rivers 2013, PIANC, Liège, Belgium/Maastricht, The Netherlands, pp 1871–1879

17. Biancardi CG (1993) Integrating ship manoeuvrability with safety. In: International conference on marine simulation and ship maneuverability (MARSIM'93), St. John's, Newfoundland, Canada, vol 1, pp 33–47
18. Alte R, Baur MV (1986) Propulsion Handbuch der Werften (in German). Hansa 18:132
19. Hochkirch K, Bertram V (2010) Engineering options for more fuel efficient ships. In: 1st international symposium on fishing vessel energy efficiency, Vigo, Spain
20. Molland AF, Turnock SR (2007) Marine rudders and control surfaces: principles, data, design and applications, 1st edn. Elsevier, Butterworth-Heinemann, Oxford

Chapter 2
Design and Evaluation of Ship Rudders

This chapter elaborates on the factors that should be considered to design a rudder. Accordingly, how to evaluate the performance of a rudder is introduced. Section 2.1 gives an introduction of the study on ship rudders. To achieve reasonable results, experiments and simulations of rudders should be carried out at relevant Reynolds numbers and angles of attack, which are discussed in Sect. 2.2. The rudder induced maneuvering forces and moments are affected by, including, but not limited to, the rudder profile, the rudder parameters, the rudder type, the number of rudders, the spacing between rudders, and the relative positions among the hull, the propeller, and the rudder. The key factors are identified through a review of each of the above-mentioned impacts on the rudder hydrodynamics in Sect. 2.3, which are further studied through CFD simulations in Chap. 4. To judge the quality of the rudder design, Sect. 2.5 presents the evaluation perspectives of the rudder performance in ship maneuverability, fuel consumption, and rudder cavitation. Section 2.6 draws a summary of this chapter.

2.1 Introduction to Ship Rudders

Rudders are primarily applied on conventional ships for course keeping and maneuvering. They are also used alone or with fins for roll stabilization [68, 103, 108, 115, 116, 122, 123, 135, 136]. In over 60 years of research on ship rudders, only a limited number of fundamental studies have been published [13, 92, 127, 142]. These studies provided valuable insights into the relationship between rudder design and rudder performance. However, it should be noted that previous studies of rudders in the open literature were mainly carried out for single NACA rudder cases, excluding the propeller and hull influences. In the past, the impacts of the rudder design on ship maneuverability and fuel consumption have not been given great attention by the researchers.

© Springer Nature Switzerland AG 2020
J. Liu, *Mathematical Modeling of Inland Vessel Maneuverability Considering Rudder Hydrodynamics*, https://doi.org/10.1007/978-3-030-47475-1_2

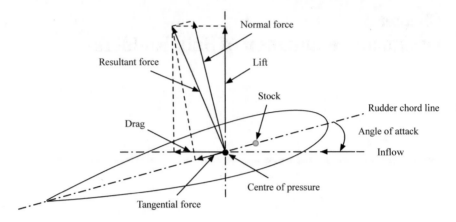

Fig. 2.1 Rudder induced forces. Adapted from Molland and Turnock [92, p. 73]

Initially, research on rudders is based on physical experiments in wind tunnels, towing tanks, or cavitation tunnels. However, these experiments have certain drawbacks, such as time and money consuming, limited model sizes, constrained Reynolds numbers, and influences of test equipment. With the rapid development of Computational Fluids Dynamics (CFD) methods, numerical analysis becomes possible. Even though CFD methods have numerical uncertainties, lack of validation data, and may also cost a lot of time and money, they still provide new opportunities for research on the rudder design. With the evolution of experiment techniques, the rudder design has been changing slowly. Therefore, the objective of this study is to gather information for researchers as well as designers to perform further developments and make proper choices.

To obtain reasonable results, the rudder hydrodynamic characteristics should be analyzed in practical conditions. This section inspects two key elements of rudder working conditions, i.e. the Reynolds number (Re) and the angle of attack (α), to set up proper test cases for the CFD simulations in Chap. 4. Under certain working conditions, rudders generate lift (F_L) and drag (F_D), based on which normal force (F_N) and tangential force (F_T) can be calculated according to the rudder angle (δ) or the angle of attack (α). Figure 2.1 illustrates the applied terminology of rudder forces in this monograph.

Lift is mainly generated by the pressure difference between the two surfaces. It is the component of the resultant force that acts perpendicular to the inflow direction. Lift nearly increases as a linear function of the angle of attack before the stall angle, which is the critical angle of attack at which maximum lift occurs. Normally, stall angles of rudders in open water are in the range of 15° to 20°, and the practical stall angles in the propeller slipstream would be larger. Furthermore, the stall angle is also affected by the rudder aspect ratio and other parameters. In astern conditions, the lift curve slope reduces to 75–85% of the ahead condition, the stall angle decreases by 5°–10°, and the maximum lift coefficient is about 45–75% of the ahead case [92, p. 99].

Drag is the rudder force component along the incidence flow direction, which consists of skin friction drag and form drag. The friction drag is caused by the frictional shear stress and determined by the size of the wet surface. The form drag, also know as viscous pressure drag or pressure drag primarily depends on the shape of the rudder. The friction drag is almost the same for rudders with the same wet area while the form drag is relatively small. At a Re of 2.4×10^5, Reichel [107] concluded that under small rudder angles (up to 5°) almost all the tested six types of rudders with the same lateral area but different profiles and constructions have the same drag coefficients.

Routinely, non-dimensional coefficients are used to compare the rudder hydro-dynamic performance with various design choices. Two main hydrodynamic characteristics are the lift coefficient (C_L) and the drag coefficient (C_D) based on which the stall angle, the maximum lift coefficient, the normal force coefficient (C_N), the tangential force coefficient (C_T), and the lift to drag ratio (C_L/C_D) are identified. These parameters are non-dimensionalized and calculated as follows:

$$
\begin{aligned}
C_L &= F_L/(0.5\rho V_R^2 A_R) \\
C_D &= F_D/(0.5\rho V_R^2 A_R) \\
C_N &= C_L \cos \alpha + C_D \sin \alpha \\
C_T &= C_D \cos \alpha - C_L \sin \alpha,
\end{aligned}
\tag{2.1}
$$

where ρ is the water density, α is the angle of attack, V_R is the rudder inflow speed, A_R is the rudder area. Whicker and Fehlner [142], Abbott and Von Doenhoff [1], Thieme [126], Molland and Turnock [92] provided further information about these coefficients.

2.2 Rudder Working Conditions

Rudder hydrodynamic characteristics indicate the rudder performance under specified conditions. The rudder working conditions determine the actual amount and direction of forces induced by the rudder. This section presents the main operational conditions, i.e. Reynolds numbers (Re) and angles of attack (α).

2.2.1 Reynolds Numbers

The Reynolds number (Re) is the ratio of the momentum force to the viscous force of flow, expressing the relative importance of these two types of forces. Rudders may have different hydrodynamic characteristics at a low Re (laminar flow) and a high

Re (turbulent flow). Based on the chord length (C_R), the full-scale *Re* of a rudder range from about 5×10^5 for a small yacht up to about 1×10^8 for a large fast ship [92, p. 34]. For the complete dynamic similarity of the flow conditions, the *Re* has to be the same for both model-scale and full-scale in experimental and numerical tests. However, a *Re* as high as 1×10^8 is commonly not achievable in contemporary physical model test facilities.

In practice, the model-scale *Re* is one order smaller than the corresponding full-scale *Re*. The difference in *Re* violates the similarity of the rudder force in model-scale and full-scale maneuvering motions. Accordingly, Ueno et al. [134], Ueno and Tsukada [133] proposed corrections of the rudder effectiveness and the speed to transfer model-scale results to full-scale. In general, an increase in *Re* leads to an increase in the lift coefficient and a decrease in the drag coefficient [74]. Furthermore, the drag coefficient is more sensitive to the change of *Re* than the lift coefficient [74]. Whicker and Fehlner [142] indicated that the *Re* influences are significant in the range of 1×10^6 to 3×10^6 and insignificant above 3×10^6. Ladson [65] noted little variations of the coefficients above a *Re* of 6×10^6. Detailed information about the impacts of Reynolds numbers on rudder hydrodynamics was given by Loftin and Smith [76].

To obtain insights into the rudder performance at high *Re* from actual low *Re* physical experiments, roughness strips are commonly attached near the model leading edge to generate turbulent flow instead of laminar flow. In practice, rudders nearly always work in fully turbulent flow due to the propeller rotation. CFD methods can study the full-scale open water rudder hydrodynamics at a high *Re*. Fach and Bertram [26] reported the progress of CFD applications in simulations of rudder flows. However, the required simulation time and computation resource increase dramatically with the model size. The impacts of Reynolds numbers on rudder hydrodynamics are further discussed in Sect. 4.2.1.

2.2.2 Angles of Attack

The angle of attack (α) or the effective rudder angle (α_R) significantly affects the amount and the direction of the rudder induced forces. α_R is closely related to the rudder angle (δ) and the ship drift angle (β), which is commonly expressed as $\alpha_R = \delta - \beta$ [144]. Greitsch [35], Greitsch et al. [36] applied operation profiles, including the frequency distributions of rudder angles and ship speeds in rudder shape optimization, cavitation risk analysis, and ship design. Brix [13] indicated that the rotation caused by the propeller may induce an incidence of $10°–15°$ to the rudder and a 10% increase in the rudder resistance. Therefore, it is beneficial to have the maximum lift to drag ratio and the minimal drag coefficient within this range to reduce the fuel consumption.

Records of rudder angles were taken on a 110 m long inland vessel by students supervised by the author. Figure 2.2 presents the probability distributions of rudder angles during the one journey from Antwerp, Belgium to Vlaardingen, the Nether-

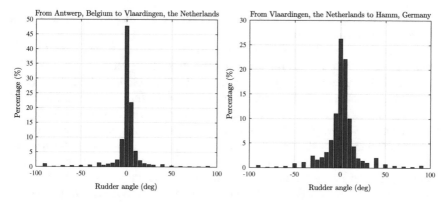

Fig. 2.2 Probability distributions of the applied rudder angles

lands and the other journey from Vlaardingen, the Netherlands to Hamm, Germany.
It shows that the most frequently used rudder angles are in the range of $-15°$ to
$15°$. This finding is quite similar to the frequency distributions published by Greitsch
[35], Greitsch et al. [36], which are rudder angles used by a ferry in the North Sea.

In Fig. 2.2, large rudder angles up to $90°$ also applied at relatively slow speed.
These operations are typical for inland vessels in hard maneuvering situations but
rarely seen on seagoing ships which have a customary maximum rudder angle of
$35°$. Inland vessels, which commonly do not have assistance for birthing or hard
maneuvering, apply the rudders at $90°$ with bow thrusters to crab or turn on the spot.
However, few studies have been made on the rudder performance at large angles of
attack. Since this research is intended for primary ship design at service speed, the
maneuvers with large rudder angles are not explicitly studied. Further research is
suggested for slow-speed maneuvering modeling with related parameters to study
the performance of crabbing, birthing, and turning with large rudder angles.

2.2.3 Concluding Remarks of Rudder Working Conditions

As discussed in the previous sections, the rudder performance depends on its working
conditions, including, but not limited to, the Reynolds number and the angle of attack.
Thus, ship operation profiles should be considered in the process of the rudder design.
A high Reynolds number can be achieved in tests by either enlarging the model size
or increasing the inflow speed. Presently, model tests at high Reynolds number or
full-scale ship tests may not be practical for primary studies due to the capacity of test
facilities or high expense. Another possible approach is the high-Reynolds-number
CFD study. Considering the cost of high Reynolds number simulations with either
model tests or CFD simulations, it is recommended to carry out tests at sufficiently
high Reynolds numbers, for instance, a Reynolds number of 6×10^6, above which

the rudder hydrodynamic characteristics may not be significantly affected by the change of the Reynolds number.

The main range of the applied rudder angles is $-35°$–$35°$, which should be the main region of interest in the rudder hydrodynamics. Histograms of applied rudder angles and ship speeds like Fig. 2.2 and results given by Greitsch [35], Greitsch et al. [36] are valuable for further studies on ship maneuvering performance and fuel consumption. More operational data of different ships with various rudder configurations in inland waterways or open sea are needed. Furthermore, research works on the rudder performance for slow-speed ships, i.e. at Reynolds numbers lower than usual, with uncustomary large rudder angles are suggested, especially for inland vessels. In summary, further investigation and experimentation into the rudder hydrodynamic characteristics at various Reynolds numbers with practical operational profiles are recommended.

Considering the Re effects, it is suggested to test a model as large as possible at a sufficiently but not extensively large Re to have a balance of computational cost and accurate results. Therefore, in this monograph, CFD simulations are performed at Re of 6×10^6 above which lift and drag coefficients are not significantly affected. For the angles of attack, this monograph presents CFD simulations with angles of attack in the range of $0°$–$35°$ to have a full vision of the tendency of the rudder hydrodynamic coefficients. maneuvering with large rudder angles are important but not studied in this monograph as it is only performed at very low speed and commonly with bow thrusters.

2.3 Rudder Design Choices

A rudder design needs to consider aspects, including, but not limited to, working conditions (Reynolds numbers and angles of attack), profiles (sectional shapes), parameters (area, thickness, span, chord, and aspect ratios), types (the position of the stock and the structural rudder-hull connection), and interactions (among the hull, the propeller, and the rudder). These design choices determine the amount and the direction of rudder forces, resulting in different rudder performance. This section presents the state-of-the-art of studies on these choices.

2.3.1 Rudder Profiles

Rudder profiles are rudder sectional shapes. The profile is usually described by the rudder camber, the position of the maximum camber, the rudder thickness, the position of the maximum thickness, and the nose radius. Figure 2.3 illustrates the terminology to be used later. Various distributions of the camber and the thickness formulate different rudder series or families.

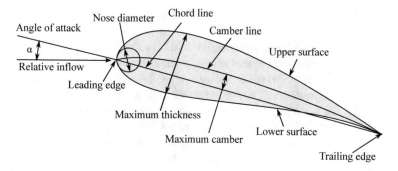

Fig. 2.3 Rudder profile terminology. Adapted from Cleynen [20]

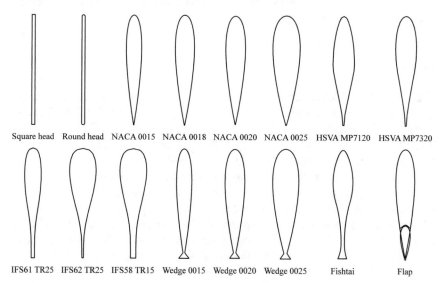

Fig. 2.4 Common rudder profiles

Typical rudder profiles applied in practice are presented in Fig. 2.4. Most of the existing rudder profiles are originally designed for airfoils like the NACA series. There are also profiles designed particularly for ship rudders such as the IFS and HSVA series [126]. Other profiles include flat-plate, wedge-tail, fishtail, and flap rudders. Various rudder profiles have different hydrodynamic characteristics, leading to a different performance in ship maneuverability, fuel consumption, and rudder cavitation.

2.3.1.1 Flat-Plate

Flat-plate profiles are normally rectangles in 2D (square head in Fig. 2.4). They are the simplest flow deflecting devices to design, the easiest profiles to produce, and the cheapest rudders to buy. To reduce the form drag, flat-plate profiles may have semi-circle or triangular leading and trailing edges with fairing tips (round head in Fig. 2.4). However, the flat-plate profile may encounter stiffness issues when a large area is required, resulting in limitations of applications. Thieme [126] indicated that the flat-plate profiles may achieve high efficiency in straight-ahead condition. However, this high efficiency only appears at small angles of attack, up to approximately 5°, after which the flat-plate profile stalls and its lift-to-drag ratio collapses. At present, flat-plate rudders are frequently seen on small boats and antique inland vessels but not common for modern seagoing ships.

2.3.1.2 NACA

NACA profiles, which are developed by National Advisory Committee for Aeronautics (NACA), are the most widely applied rudder profiles [60] (Fig. 2.4). They are also applied to other foil-shaped structures such as, propellers [125], propeller ducts [62, 145], marine current turbines [6, 7, 33, 93], fins [106, 122, 123], and wind turbines [9, 128]. Furthermore, the NACA series is the most thoroughly investigated airfoil family. Therefore, it is commonly taken as a benchmark case for both aerodynamic and hydrodynamic studies. The geometry of the NACA series was described by Ladson et al. [64]. Wind tunnel test results for airfoils at small Mach numbers, which means the air is almost incompressible like water, are applicable for ship rudders [1, 34, 65, 76, 79]. Especially, characteristics of airfoils with low aspect ratios [29, 102, 111, 129] are quite close to those of common ship rudders. Whicker and Fehlner [142], Thieme [126] further discussed applications of the NACA profiles to shipping. In general, NACA profiles can generate sufficient maneuvering force with high efficiency.

2.3.1.3 HSVA

HSVA profiles were specially developed for ship rudders by Hamburg Ship Model Basin (Hamburgische Schiffbau Versuchsanstalt GmbH, Hamburg, Germany). Considering the rudder working conditions, the HSVA series is designed to have a good pressure distribution that reduces the onset of cavitation [63]. Bertram [10, p. 271] provided offsets of two main HSVA profiles, namely HSVA MP7120 and HSVA MP7320. Thieme [126], Brix [13] presented hydrodynamic characteristics of the HSVA profiles in detail. According to Hollenbach and Friesch [45], high-lift HSVA profiles may reduce the rudder area, achieving 1% fuel saving.

2.3.1.4 IFS

IFS profiles were developed by Institute für Schiffbau, Hamburg, Germany to achieve a steep lift curve slope, a large stall angle, and a high maximum lift coefficient, as shown in Fig. 2.4. Bertram [10, p. 271] showed the offsets of three main IFS profiles, which are IFS58 TR15, IFS61 TR25, and IFS62 TR25. Thieme [126] presented wind tunnel tests of the IFS profiles. Compared to the HSVA profiles, IFS profiles may generate slightly more lift, induce more drag, and suffer less cavitation [10, p. 297].

2.3.1.5 Wedge-Tail

Wedge-tail profiles are simplified fishtail profiles (Fig. 2.4), which normally have a sharp concave point. Van Nguyen and Ikeda [140] indicated that the hydrodynamic characteristics of wedge-tail rudders are related to the size of the tail and the profile thickness. Through CFD simulations, Liu and Hekkenberg [73] presented that the tested wedge-tail profiles can generate more lift than the tested flat-plate and NACA profiles at the cost of additional drag. Since no standard offsets of wedge-tail profiles are found, it is hard to compare the performance of wedge-tail rudders in literature.

2.3.1.6 Fishtail

Fishtail profiles (Fig. 2.4), also know as Schilling rudders [112, 113], are normally developed based on conventional NACA, HSVA, and IFS profiles with trailing tails. The concave part, where the original profile connects with the tail, is smoothed to have a better pressure distribution that delays stalling. Fishtail rudders can effectively generate lift, improving the ship maneuverability. Therefore, they are frequently used in inland vessels. Van Nguyen and Ikeda [139], Van Nguyen and Ikeda [138] developed high-lift fishtail profiles by optimizing the maximum rudder thickness and the trailing edge thickness. Hasegawa et al. [42], Nagarajan et al. [95] discussed the superiority of the fishtail rudder to the conventional Mariner rudder of the course keeping ability in windy conditions. However, very few studies have examined the fishtail profiles by experimental tests. In addition, offsets of the fishtail profiles are not publicly available.

2.3.1.7 Flap

A flap profile has a movable flap which changes the profile camber (Fig. 2.4). Therefore, flap profiles can improve ship maneuvering performance without significantly affecting its cruising characteristics. The disadvantages of flap rudders are large hinge moments, high mechanical complexity, and potential maintenance difficulties [100]. Two main parameters of a flap rudder are the flap-linkage ratio (the flap angle relative to the rudder chord line divided by the rudder angle) and the flap-area ratio (the sectional area of the flap divided by the total sectional area). Olson [99] indicated that an increase in either the flap-linkage ratio or the flap-area ratio increases the lift coefficient, reduces the rudder efficiency, and shifts the center of pressure to the rear in ahead condition while for the astern condition, the lift coefficient is decreased.

Bertram [10, p. 284] described that flap rudders may provide a much higher lift curve slope and 60–70% higher maximum lift compared to a conventional rudder of the same shape, size, and area. Olson [99] reported that a 30% flap NACA 0018 profile using a 1.5 flap-linkage ratio can generate 50% higher lift than an all-movable rudder of equal area. Kerwin et al. [52–54], Oppenheim [100] indicated that a 20% flap NACA 66 profile may achieve 59% higher lift than the original NACA 66 profile. Kerwin et al. [53] showed that the drag coefficient at zero lift increases with the flap size and concluded that even a small flap can significantly increase the maximum lift coefficient. Additionally, the size of the flap in a range of 20–50% of the total rudder area does not influence the maximum lift coefficient [53].

However, the increase in maximum lift is achieved at the expense of a large increase in the drag and the hinge moment [53, 54]. The flap balance may reduce the flap moment but it also significantly reduces the rudder induced side force [52]. Thus, Kerwin et al. [52] suggested that a rudder with a small, unbalanced flap might achieve a balance of the improvement over the all-movable rudder and the practical structural requirement. Champlain [17] analyzed the effects of the flap gap (the distance between the trailing edge of the skeg and the leading edge of the flap with zero flap deflection) on rudder hydrodynamic characteristics indicating that a larger open gap may achieve better overall performance.

2.3.2   Rudder Parameters

After a selection of the profile, detailed design of rudder parameters should be considered. To avoid ambiguity, the rudder parameters are illustrated in Fig. 2.5. These parameters affect the rudder hydrodynamic characteristics and determine the rudder effectiveness and efficiency. Main parameters discussed in this monograph are the area, the thickness, the span, the chord length, and the aspect ratios. These parameters are determined as the key parameters and examined in Chap. 4.

Sweep angles, taper ratios, and tip shapes have relatively small influences on the rudder hydrodynamic characteristics [92], therefore, they are only briefly introduced here. The sweep angle slightly affects the maximum lift coefficient and the stall angle [91, p. 79]. An increase in the taper ratio leads to an increase in the lift curve slope, the maximum lift coefficient, and the stall angle [92, p. 90]. A fairing tip shape may reduce the minimum drag at zero angle of attack but it decreases the stall angle by 2°–3° [142]. Hoerner [44], Molland and Turnock [92] concluded that the square tip is better than the fairing tip because the small advantage of the fairing tip in the reduction of drag at small rudder angles is gained at the expense of large hydrodynamic losses at large rudder angles.

Fig. 2.5 Rudder parameter terminology. Adapted from Molland and Turnock [92, p. 72]

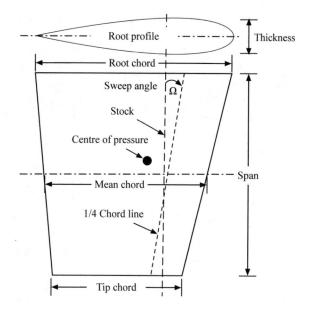

2.3.2.1 Area

The rudder area (A_R) affects the amount of lift and drag induced by the rudder. To generate required maneuvering forces and moments, rudders (one or more) should have a sufficient total area. Multiple rudders are applied when the area of a single rudder is insufficient to turn the ship at the required rate. One reason why the rudder cannot have the required area is because the draft of the ship is small, for instance, inland vessels. However, a larger rudder area normally means larger rudder induced resistance.

The rudder area or the total rudder area, if more than one rudder is applied, is commonly expressed as a ratio of the ship lateral underwater area (LT), where L is the ship length between perpendiculars and T is the loaded ship draft. The value of A_R/LT is normally first estimated based on similar ships or empirical formulas, and then optimized through iterations [60]. Table 2.1 summaries the reference A_R/LT values found in literature [5, 60, 92].

Reference values in Table 2.1 indicate that the ships which have high requirements of maneuverability, such as warships, pilot vessels, tugs, and trawlers, need a large rudder area. For rudders working directly behind propellers, Veritas [141] suggested that A_R/LT should not be less than:

$$\frac{A_R}{LT} = 0.01 \left[1 + 50C_b^2 \left(\frac{B}{L} \right)^2 \right],$$

(2.2)

Table 2.1 Reference ratios of the rudder area to the ship lateral underwater area for different ship types

Ship types	A_R/LT (%)
Container ships	1.2–1.7 [5, p. 88]
Passenger liners	1.2–1.7 [5, 60, p. 88]
General cargo ships	1.5 [5, p. 88]
Single-propeller merchant ships	1.6–1.8 [92, p. 189]
Twin-propeller merchant ships	1.6–2.2 [92, p. 189]
Small cargo ships	1.7–2.3 [60]
Large cargo ships	2.0–2.8 [60]
Oil tankers and bulk carriers	2.0 [5, p. 88]
Lake steamers	2.0 [5, p. 88]
Cross channel ferries (RoRo ships)	2.0–3.0 [5, p. 88]
Coastal vessels	2.0–3.3 [5, p. 88]
Warships	2.4–2.8 [92, p. 189]
Pilot vessels	2.5–4.0 [5, p. 88]
Tugs	3.0–4.0 [92, p. 189]
Trawlers	3.0–4.0 [92, p. 189]
Inland cargo vessels in non-rapid flow segment	2.0–3.0 [18, p. 11]
Inland cargo vessels in rapid flow segment	4.5–5.0 [18, p. 11]

where C_b is the ship block coefficient. Additional 30% area should be added if the rudder does not work directly behind a propeller [141].

To ensure a quick response to helm for wide ships, Schneekluth and Bertram [114, p. 62] advised relating the rudder area (A_R) to the ship midship section area (A_m) and A_R should not be less than 12% of A_m. To satisfy a particular turning index, Clarke et al. [19] reported that as C_b increases A_R increases slightly while as B/L increases A_R increases significantly, especially above B/T larger than 3. Thus, inland vessels which typically have larger block coefficients, much larger L/B ratios, and extremely much larger B/T ratios than seagoing ships [105] should carefully consider the size of each rudder and the number of rudders to have sufficient total area.

2.3.2.2   Thickness

The rudder thickness is commonly expressed as a ratio of the rudder chord length. It needs to satisfy the structural needs and, furthermore, affects the minimum drag, the stall angle, and the maximum lift coefficient [92, p. 92]. Commonly, thinner profiles have higher efficiency, more specifically a higher lift to drag ratio, than thicker profiles. Van Beek [137] indicated that a slim rudder profile may increase the propulsive efficiency by 1–3%. A rudder may have a span-wise different thickness, for instance, a thin profile at the tip and linearly increases to a thick rudder profile at the root. This configuration may provide a balance of the structural requirement and the hydrodynamic efficiency.

2.3.2.3 Span and Chord Length

The rudder span or rudder height (B_R) is the distance between the rudder root and tip sections. Normally, the span is expected to be as large as possible, which may ensure a large geometric rudder aspect ratio for high effectiveness and efficiency. However, the span is commonly constrained by ship particulars (ship draft) and operational profiles (water depth).

From observation and experience, for inland vessels, a normal value of the rudder span is around the size of the propeller diameter. The rudder chord length (C_R) is the distance between the leading and trailing edges. The chord length is commonly determined according to the rudder area, the geometric aspect ratio, and the span. For unbalanced rudders, a large chord length is not favorable as it may put an excessive burden on the steering gear.

2.3.2.4 Geometric and Effective Aspect Ratios

Rudder aspect ratios have the most significant influence on rudder hydrodynamic characteristics [92, p. 89]. It includes two concepts: the geometrical aspect ratio (Λ_G) and the effective aspect ratio (Λ_E). Λ_G is commonly expressed as $\Lambda_G = B_R/C_R$ or $\Lambda_G = B_R^2/A_R$. The effective aspect ratio (Λ_E) is the actual aspect ratio applied to hydrodynamic force calculation.

In general, a rudder with a larger geometric aspect ratio can generate a larger lift with lower drag [142]. However, a small geometric aspect ratio may enhance the maneuverability with a large stall angle [92, p. 64]. Confirmed by a rudder manufacturer, a common range of geometric aspect ratios for seagoing ships is 1.5–3.0 while the common range of aspect ratios for inland vessels is 1.0–2.0.

To compare the hydrodynamic characteristics of profiles, an infinite geometric aspect ratio (an infinite span), is commonly assumed in experimental and numerical studies [39, 79]. Such an approach associates the hydrodynamics only with the parameters of the 2D profile rather than the 3D shape of the rudder. Thus, it is useful for investigating the profile drag, the pressure distribution, and theoretical application [92, p. 41].

In practice, the rudder has a finite span and, thus, a finite aspect ratio. A finite span has a downward flow along and behind the rudder, which is the so-called downwash. This downwash is combined with the inflow leading to a smaller effective angle of attack than the deflected rudder angle. Therefore, small geometric aspect ratios have a larger reduction of effective rudder angles due to larger downwash leading to a larger stall angle.

The effective aspect ratio (Λ_E) is commonly estimated based on the geometric aspect ratio (Λ_G). When the rudder root is sufficiently close to a flat surface, such as

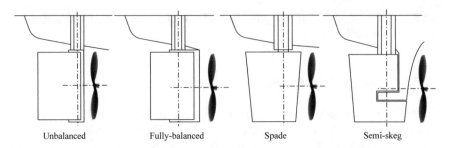

Fig. 2.6 Common rudder types. Adapted from Molland and Turnock [92, p. 15]

a large endplate or a flat hull, Λ_E/Λ_G is close to 2 owing to the ideal mirror image effect [92, p. 183]. Considering the gap effects, Brix [13, p. 97] provided reference values of Λ_E/Λ_G. Root and tip endplates may increase the effective aspect ratio and cause notable drag. From observation, seagoing ship rudders normally do not have endplates while inland vessel rudders tend to have both root and tip plates.

2.3.3 Rudder Types

Rudder types are classified based on two aspects: the position of the stock (unbalanced, semi-balanced, or balanced) and the structural rudder-hull connection (the number of pintles, without a skeg, semi-skeg, or full-skeg). The choice of the rudder type depends on the ship type, the ship main dimensions, the shape of the stern, and the required rudder size [92, p. 13]. This section reviews four common rudder types, namely unbalanced, fully-balanced, spade, and semi-skeg, which are shown in Fig. 2.6.

2.3.3.1 Unbalanced

Unbalanced rudders have their stocks at the leading edge, as a result of which, the entire rudder area is located after their stocks (Fig. 2.6). The steering gear has to provide all the rudder turning torque. It implies that this solution only works for rudders with a limited area. Otherwise, the rudder cannot be steered properly. To compensate for the large bending moment, this type of rudder has two pintles on top and bottom. Currently, unbalanced rudders are not widely used for modern single-propeller ships, but they are still popular for small crafts and fishing vessels as the unbalanced rudders are easy and cheap to produce [10, p. 282].

2.3.3.2 Fully-Balanced

Fully-balanced rudders have their rudder stocks at 20–40% chord length from the leading edge (Fig. 2.6). The water force acting on the aft part of the rudder is partially or, at some rudder angles, completely compensated by the force acting on the forward part of the rudder. Therefore, the rudder turning torque and the required capacity of the steering gear for fully-balanced rudders are far less than that for unbalanced rudders. Since the action point of the force changes with the rudder angle, it is not feasible to maintain the balance over a complete range of rudder angles. Fully-balanced rudders are extensively applied to single-propeller merchant ships and gradually superseded by semi-balanced skeg rudders [92, p. 14].

2.3.3.3 Semi-skeg

Semi-skeg rudders, also called horn rudders, or Mariner rudders, are semi-balanced rudders, more specifically unbalanced root part with a skeg and balanced tip part without a skeg (Fig. 2.6). The location of the pintles is supposed to be in the vicinity of the center of pressure, affecting the response and torque characteristics of the rudder. The horn provides structural support for the span-wise bearing moment, making a large rudder area possible. In addition, semi-skeg rudders require less turning torque than unbalanced rudders and have less bending moment than spade rudders. Nowadays, semi-skeg rudders tend to be favorable for newly built large ships despite the hydrodynamic advantages of spade rudders [77].

Through series of free-stream wind-tunnel tests, Molland [82, 83] found that a semi-skeg rudder has a little smaller maximum lift, a smaller lift to drag ratio, and a significantly smaller lift curve slope than an all-movable rudder of the same size. Thus, semi-skeg rudders are less effective than spade rudders for maneuvering [10, p. 283]. These changes in lift and drag coefficients are mainly caused by the rudder horn. Even though the rudder horn itself does not incline, it still significantly affects the lift and drag of the rudder [77].

2.3.3.4 Spade

Spade rudders are balanced rudders with a taper ratio (the ratio of the root chord to the tip chord) as shown in Fig. 2.6. Rudders with a large taper ratio may reduce the rudder drag or even generate thrust [16]. Due to the large bending moment, spade rudders are commonly designed with a large stock diameter and a large rudder thickness. According to Bertram [10, p. 283], spade rudders are not feasible if the required stock diameter is larger than 1 m. Commonly, spade rudders consume less energy to be operated than the unbalanced rudders.

Bertram [10, p. 272] stated that spade rudders are more favorable than unbalanced or fully-balanced rudders. From hydrodynamic and cavitation points of view, spade rudders are better than semi-skeg rudders. At an optimum relative position to the propeller (about 30–35% of the propeller diameter), a spade rudder shows about 1.6% gain of power against a semi-skeg rudder [81]. Unlike semi-skeg or full-skeg rudders, spade rudders do not have gap cavitation, reducing the time and cost for maintenance. Nowadays, spade rudders are widely applied to all ship types.

2.3.4 Concluding Remarks of Rudder Design Choices

2.3.4.1 Concluding Remarks of Rudder Profiles

Through the previous discussion, it is clear that different rudder profiles have different hydrodynamic characteristics in the lift and drag coefficients, the slope of the lift curve, the stall angle, and the lift to drag ratio. Thieme [126], Molland and Turnock [92] further compared the performance of various rudder profiles. Liu et al. [72, 75] proposed regression formulas to estimate the rudder normal force of different rudder profiles and analyzed their impacts on ship maneuverability. These differences in the characteristics of the profile should be carefully considered in the rudder design. Furthermore, it is recommended to make a uniform definition of the shapes of wedge-tail and fishtail profiles to make research outcomes easier to verify and to expand upon the existing research. In general, the choice of the rudder profile depends on ship particulars, operational requirements of maneuvering performance, and fuel conservation.

For seagoing ships which sail long distance and commonly have tug assistance for hard maneuvering in the port area, the efficiency of the rudder may have a higher priority than the effectiveness. Thus, on the prerequisite of sufficient rudder force for course keeping and maneuvering, highly efficient profiles are suggested, such as NACA, HSVA, and IFS. For ships mainly sail in constrained waterways like inland vessels, the effectiveness of the rudder is more crucial than efficiency. Therefore, high-lift profiles, including fishtail, wedge-tail, and flap, are proposed. When the applicable rudder area is limited due to ship draft or water depth, high-lift profiles are also favorable. In the case of working boats like tugs, where the maneuverability of the ship is the key, the effectiveness can be the primary concern while the efficiency can be sacrificed. All things considered, the lift is nearly always gained at the expense of drag. A wise decision of the rudder profile should coincide with the objective of the design of the ship.

2.3.4.2 Concluding Remarks of Rudder Parameters

Besides the rudder profile, the rudder parameters specify how the rudder is shaped in three dimensions. The rudder performance is the end product of all these properties. First of all, the total area should be sufficiently large as it is a determinant factor of the amount of rudder force. The reference values of the total area are given in Table 2.1. When the area of a single rudder is not sufficient, multiple-rudder configurations are recommended for inland vessels because of the limited ship draft and seagoing ships due to the enlargement of the ship dimensions. In addition, the increase in the total rudder area leads to an improvement in the ship maneuverability but makes the rudder induced resistance larger. Therefore, for ships with large rudder area, the selection of the rudder profile becomes even more crucial.

As a general rule, thinner rudders have better hydrodynamic performance than thicker ones. The structural requirement and the hydrodynamic performance have to be considered in coincidence in the determination of the rudder thickness. With a prerequisite of the total rudder area, a large span means short chord, furthermore, a large geometric aspect ratio, which is desirable for both efficiency and effectiveness. The span, the chord, and the geometric aspect ratio are commonly limited in a routine range, but the effective aspect ratio can be enlarged by adding endplates on the tip and the root of the rudder. Yet, very few studies have examined the impacts of the shape and the size of the endplates on the rudder hydrodynamics. As a summary, emphasizing the ship maneuverability, large total rudder area, small thickness, large span, short chord, large geometric and effective aspect ratios are recommended.

2.3.4.3 Concluding Remarks of Rudder Types

As a summary, conventional propeller-rudder systems still hold a dominant position on merchant ships. Even though active steering devices have been developing rapidly for offshore engineering, the conventional rudders can still provide extraordinary performance regarding bollard pull, response times, and cruise behavior [71]. Following consideration of the rudder type, spade rudders and semi-skeg rudders are primarily contemporary design choices. From the perspective of hydrodynamics, spade rudders are better than semi-skeg rudders for ship maneuverability and fuel consumption. Thus, it is proposed to take the spade rudder as a first choice. However, the area of a spade rudder is limited due to the high bending moment. Therefore, semi-skeg rudders are recommended for large ships that require a large rudder area.

2.4 Interaction Effects on Rudders

The propeller accelerates and rotates the wake of the hull to the rudder. The rudder works in the propeller slipstream, inducing maneuvering forces and moments. The upstream propeller affects the rudder induced maneuvering force while the presence of the rudder influences the propeller generated thrust and the required torque [131]. These interactions affect the performance of the propeller and the rudder, as well as the total effectiveness and efficiency of the ship [81]. This section discusses the interactions among the hull, the propeller, and the rudder in four aspects: flow straightening effect, propeller slipstream, relative positions of the propeller and the rudder, and multiple rudders.

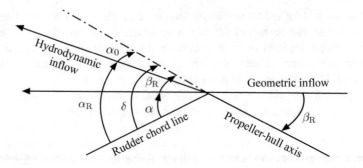

Fig. 2.7 Flow straightening terminology. Adapted from Molland and Turnock [89], Badoe [4]

2.4.1 Hull Flow Straightening Effects

When the ship is drifting at a certain drift angle (β), the effective rudder angle (α_R) decreases. However, the hull straightens the inflow into the propeller-rudder system recovering the loss of the effective rudder angle. Figure 2.7 illustrates the terminology applied in the flow straightening effect.

The relationship of the terminology is as follows:

$$\delta = \alpha + \beta_R = \alpha_R + \alpha_0 = \alpha_R + \gamma_R \beta_R, \tag{2.3}$$

where δ is the rudder angle relative to the ship centerline, α is the geometric rudder angle, α_0 is the incidence for zero lift, β_R is the drift angle at the position of the rudder which is larger than the ship drift angle β, and γ_R is the flow straightening factor. γ_R depends on the form of the hull body, the drift angle, and the propeller working load [91]. The magnitude of the drift angle effect on the rudder performance depends on the advance ratio or propeller working load. A lower advance ratio leads to a higher propeller working load and a more significant flow straightening influence. The drift angle shifts but does not greatly change the shape of the rudder force coefficient curve [88].

2.4.2 Propeller Slipstream Effects

The propeller advance ratio (J_P), which expresses the speed of the propeller slipstream compared to the ship velocity, is calculated as follows:

$$J_P = \frac{V_P}{n_P D_P}, \tag{2.4}$$

where V_P is the propeller advance speed, n_P is the propeller resolution, D_P is the propeller diameter. A high J_P (low or zero propeller thrust) indicates that the rudder performance depends on the hull, the free surface, and the ship yaw angle while a low J (high propeller thrust) indicates the rudder induced forces is determined by the propeller slipstream [131]. The propeller working load K_T/J_P^2 forms the speed and the pressure distribution of the propeller slipstream. An increase in the propeller working load increases the rudder induced side force while delays the stall angle [84, 87, 131]. Meanwhile, the presence of a rudder behind the propeller increases the propeller working load [119, p. 155].

Rudder hydrodynamic characteristics tested in open water are different from those tested in the propeller slipstream. Oppenheim [100] found that the lift curve slope, the maximum lift coefficient, the stall angle, and the drag coefficient increase when the rudder is tested in the propeller slipstream. For spade rudders, Nienhuis [98] showed that the stall angle is significantly delayed in the propeller slipstream but the slope of the lift curve is not much affected. Kerwin et al. [52] compared the rudder hydrodynamics of a 20% flap rudder in propeller slipstream and uniform flow indicating that the rudder lift curve slope is increased by about 25%.

2.4.3 Relative Positions of Rudders to Hull and Propeller

The longitudinal separation (x_{PR}) is the distance between the rudder stock and the propeller rotating plane. It determines the diameter and the velocity distribution within it of the propeller slipstream arriving at the rudder [131]. According to Molland and Turnock [84], x_{PR} has little impacts on the rudder side force. Oppenheim [100] concluded that the steady forces on the rudder are completely independent of x_{PR} in the range of $0.5D_P$ to $1.0D_P$. x_{PR} also affects the rudder induced resistance. Under some extraordinary conditions, the rudder may generate thrust reducing the overall resistance of the ship [16]. At various ship speeds, Reichel [107] tested six rudders at x_{PR} of $0.59D_P$, $0.65D_P$ and $0.71D_P$ and concluded that the best rudder location is the closest to the propeller. Minchev [81] tested spade and semi-skeg rudders at x_{PR} of $0.456D_P$, $0.371D_P$ and $0.272D_P$ and showed that the optimum x_{PR} for a single-propeller single-rudder bulk carrier could be in the range of $0.30D_P$ to $0.35D_P$.

The lateral separation (y_{PR}) is the distance between the rudder stock and the propeller shaft. When the number of propellers and the number of rudders are the same, the rudder central plane commonly aligns with the propeller shaft. A change in y_{PR} leads to a shift in the rudder incidence for zero lift while the hydrodynamic characteristics are not greatly affected [49, 85, 86, 90, 131]. In addition, this shift increases with an increase in the propeller working load [90]. The vertical separation (z_{PR}) is the distance between the rudder tip and the propeller shaft. A change of z_{PR} alters the proportion of the rudder span in the propeller slipstream leading to a shift in the incidence of zero lift. In general, the vertical separation has relative small influences on rudder performance.

Both experimental and numerical methods have been applied for the studies on the propeller and the rudder. Turnock [130] described a wind-tunnel test rig for the propeller-rudder interaction effect. This test rig has been successively applied for experimental tests [84–88, 90, 91]. Shen et al. [118], Felli et al. [28], Felli and Falchi [27], Pecoraro et al. [101] carried out Laser Doppler Velocimetry (LDV) measurements. Turnock and Wright [132], Laurens [66], Lee and Fujino [67], Natarajan [97], Phillips et al. [104], Caldas et al. [15], Di Mascio et al. [23], and Badoe et al. [4] presented CFD studies on the propeller-rudder interactions. Stuck et al. [121] performed an evaluation of the RANS method for predication of steady rudder performance. In addition, the rapid development of CFD provides new possibles of full-scale tests avoiding the challenge of scaling *Re* as mentioned in Sect. 2.2.

2.4.4 Multiple-Rudder Interactions

When multiple rudders are applied, the interaction among the rudders should be considered. Multiple-rudder ships may have different starboard-side and port-side maneuvering behaviors. The asymmetrical behaviors are notable for single-propeller twin-rudder ships [14, 22, 37, 41, 51, 96]. According to Yoshimura and Sakurai [147], hydrodynamic characteristics of a twin-propeller twin-rudder are not so much different from those of a single-propeller single-rudder ship. Quite a few studies have been done for twin-propeller twin-rudder ships [11, 21, 24, 25, 50, 55–57, 59, 69, 124, 146, 147]. Additionally, hydrodynamic characteristics of each rudder in twin-rudder configurations are also affected by the interaction between the rudders [32].

For twin-rudder ships, the distance between the rudder stocks and the coupling of the rudder angles may have significant impacts on ship maneuverability. By setting both rudders outwards at 75°, covering the gap between the leading edges, a twin-rudder ship may reduce the stopping distance by 50% compared to a conventional reverse engine stopping [8, p. 41]. These outwards rudder angles are called clam shell angles as shown by Hamamoto and Enomoto [37] and discussed by Hasegawa et al. [40].

Hamamoto and Enomoto [37] proposed analytical formulas of the ship forward speed drop, the stopping time, and the stopping distance when a ship stops with the clam shell angles. Although the above studies invested the maneuverability of some twin-rudder seagoing ships, no reference in the literature described the maneuvering performance of multiple-rudder inland vessels. In fact, inland vessels are more commonly equip twin rudders or even quadruple rudders than seagoing ships.

2.4.5 Concluding Remarks of Interaction Effects on Rudders

As a rule of thumb, the interactions among the hull, the propeller, and the rudder affect the rudder hydrodynamic characteristics. The flow straightening effect influences the effective rudder angle, which is important for the calculation of the rudder induced force in maneuvering simulations. However, in the existing literature, the flow straightening factor is primarily determined by model tests. More research is needed to better understand the roles of the impact factors on the flow straightening effect. Furthermore, it is proposed to generate regression formulas of the flow straightening factor through a series of benchmark tests.

The propeller slipstream mainly delays the stall angle. It may maintain or change the slope of the lift curve depending on the working load of the propeller. Further research in this field would be of great help in maneuvering simulations. The relative position of the propeller and the rudder influences the performance of both the propeller and the rudder. With extra consideration of cavitation, it is recommended to put the rudder as close as possible to the propeller. Moreover, interaction among multiple rudders requires further study, which is particularly meaningful for inland vessels.

Last but not least, both experimental and numerical methods are applicable for studies on the propeller and the rudder, which provide lots of possibilities for further research. It is recommended to use numerical methods for primary studies and apply experimental tests as a final check. Additionally, benchmark experimental tests are needed for the validation of the numerical methods.

2.5 Rudder Performance Evaluation

Good performance in ship maneuverability, fuel consumption, and rudder cavitation is the goal of a rudder design. Above all, rudders should be capable of inducing sufficient maneuvering forces to ensure navigation safety, especially for ships which frequently sail in constrained waterways or severe conditions (Sect. 2.5.1). The rudder should also be efficient, which means minimum drag at the required lift (Sect. 2.5.2). Last but not least, rudder cavitation should be considered to reduce the time and cost of maintenance (Sect. 2.5.3).

2.5.1 Ship Maneuvering Performance

Commonly, the effectiveness and efficiency of the rudders are evaluated by the amount of the rudder induced side force (Y_R) and the ratio of Y_R to the rudder induced resistance respectively. In open water tests, the effectiveness and the efficiency are commonly presented by the lift coefficient (C_L) and the lift to drag ratio (C_L/C_D).

For maneuvering simulations, only rudder normal force (N_R) is considered in the calculation of Y_R and rudder tangential force is neglected. Furthermore, N_R is routinely estimated based on the lift gradient coefficient expressed by the Fujii formula [30, 31] as the following:

$$f_\alpha = \frac{6.13 \Lambda_G}{\Lambda_G + 2.25}, \tag{2.5}$$

where f_α is the lift gradient coefficient and Λ_G is the rudder geometric aspect ratio.

The Fujii formula (Eq. 2.5) does not take the design choices as described in Sect. 2.3 into account. Thus, research on rudder design choices is especially needed for inland vessels that have a large range of variety of rudder configurations. Considering the characteristics of the rudder profile, Liu et al. [72, 75] proposed regression formulas for the normal force coefficients of various profiles based on CFD results. Furthermore, Liu and Hekkenberg [73] showed that each rudder of a twin-rudder configuration has different hydrodynamic coefficients. Motoki et al. [94] discussed the effects of the rudder horn and the propeller vortex on maneuvering simulations.

The rudder effectiveness in ship maneuverability is commonly evaluated by the amount of the rudder induced side force (Y_R), which is the component of the rudder resultant force normal to the ship centerline [13, p. 96]. Y_R can be calculated as the following:

$$Y_R = 0.5 \rho V_R^2 A_R C_{YR}, \tag{2.6}$$

where C_{YR} is the rudder side force coefficient and is normally estimated based on the gradient of the side force coefficient C_{YR_α} as $C_{YR} = \alpha C_{YR_\alpha}$. Normally, C_{YR_α} is available from experimental results or empirical formulas as the following:

$$C_{YR_\alpha} = \frac{1.8\pi \Lambda_E}{\sqrt{\Lambda_E^2 + 4} + 1.8}, \text{ by Mandel [78]}, \tag{2.7}$$

$$C_{YR_\alpha} = \frac{2\pi \Lambda_E(\Lambda_E + 1)}{(\Lambda_E + 2)^2}, \text{ by Söding [120]}, \tag{2.8}$$

where Λ_E is the rudder effective aspect ratio. More frequently, the rudder side force is calculated by the rudder normal force (F_N), neglecting the rudder tangential force (F_T) [58, 144] as the following:

$$Y_R = -(1 + a_H)F_N \cos \delta, \tag{2.9}$$

where a_H is the rudder force increase factor due to the hull. The rudder normal force (F_N) is expressed as:

$$F_N = 0.5 \rho V_R^2 A_R f_\alpha \sin \alpha, \tag{2.10}$$

where $f_\alpha \sin \alpha$ stands for the rudder normal force coefficient (C_N). According to Bertram [10, p. 272], the rudder effectiveness can be improved through the following methods:

- Increase the percentage of the rudder surface in the propeller slipstream.
- Increase the total rudder area.
- Apply more suitable rudder types (e.g. the spade rudder instead of the semi-skeg rudder).
- Apply more powerful rudder steering engines allowing larger rudder angles than the customary maximum rudder angle of 35°.
- Higher rudder turning speed.

2.5.2 Fuel Consumption Performance

A rudder may increase total resistance by about 1% in the neutral position and 2–6% at moderate angles [3]. Aiming to cut CO_2 emissions, International Maritime Organization [47, 48] requires that all ships larger than 400 gross tonnage reduce Energy Efficiency Design Index (EEDI) by up to 30% after 2025. To achieve such a goal, more efficient rudders, which can induce sufficient lift with minimum drag, are helpful. Furthermore, minimized rudder torque can also reduce the fuel consumed by the steering gear. In general, 2–8% saving can be achieved by optimizing the rudder configuration in terms of profiles, types, and efficiency-improving devices [43].

Hochkirch and Bertram [43] pointed out that the rudder has an underestimated potential for fuel reduction, for instance, reducing the rudder size (weight and resistance) by improving the rudder profile or changing to a highly efficient flap rudder. Lehmann [70] summarized that an efficient rudder system should have a slim and low drag rudder profile, generate high lift at small rudder angles, have a smooth surface, be tuned with the propeller, be light-weighted, and be easy to maintain. Hollenbach and Friesch [45] listed possible maximum gains of fuel reduction by optimizing the arrangement and shape of the propeller-rudder system.

Lehmann [70] suggested optimizing the propeller-rudder system and reducing the rudder weight to save fuel. Lehmann [70] indicated that it is important to integrate rudder design with propeller and hull form design. Van Beek [137], Lehmann [70] applied a torpedo-shaped bulb on the rudder as a streamlined continuation of the propeller hub. Similarly, Hollenbach and Reinholz [46] found that the ships with a bulb fitted rudder require 4% less power than those with the standard rudder. Sarasquete et al. [110] showed a 12% reduction of power demand for a fishing vessel by modifying the propeller hub and rudder shapes.

Reducing drag due to rotating incidence flow, a twisted rudder may enhance the overall propulsive efficiency. Commonly for a clockwise rotating propeller, the leading edge of the rudder above the shaft center is twisted port and below the shaft is twisted starboard. A twisted rudder with a Costa bulb may have 4% less fuel consumption [45]. Kim et al. [61] reported that a Z-twisted rudder, which has a

Fig. 2.8 Typical areas of
cavitation damage on a
semi-skeg rudder. Adapted
from Rhee et al. [109]

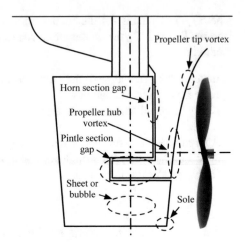

Z-shape leading edge, with and without a fin may reduce the fuel consumption by
2.35 and 2.95% respectively. Due to a decrease in the effective angle of attack, the
lift and drag of twisted rudders may be smaller than those of common spade rudders
[61]. Yang et al. [143] further studied the rudder forces of a twisted rudder.

2.5.3   Rudder Cavitation Performance

Rudders are placed in high-speed propeller slipstream. Cavitation happens when the
pressure in the flow is as low as the water vapor pressure. Brennen [12, Sect. 3.6]
explained that the cavitation damage is caused by the repetition of the cavitation
bubble collapse in the vicinity of a solid surface. This collapse generates highly
localized and transient surface stresses, which causes local surface fatigue failure and
eventually develops to erosion. Due to repair or replacement of the eroded rudder,
maintenance cost increases and operational time decreases [117]. Meanwhile, rudder
cavitation also causes an increase in drag, hull vibration, and radiated noise [117].

The ship enlargement and its impacts on ship speed lead to higher speed and
lower pressure in the propeller slipstream. Rudder cavitation has become more and
more serious [2, 38] in the past years. However, the service speed of ships tends
to be decreased to save fuel and meet the EEDI requirements [43, 45, 70]. Thus,
cavitation may become less significant in the future. As Rhee et al. [109] showed,
Fig. 2.8 illustrates typical areas of cavitation damage on a semi-skeg rudder. The
damages are mainly due to the high speed near the horn and pintle section gaps, the
propeller tip and hub vortex, and the propeller sheet.

Cavitation causes cavity drag. The cavity drag increases sharply with an increase
in ship speed [117]. Thus, a reduction of ship resistance is expected if the rudder can
be operated without cavitation, especially for high-speed vessels. Lübke [77] showed

that the cavitation effects cause a 10% decrease in lift and a 20% increase in drag. According to Shen et al. [118], Mewis and Klug [80], Ahn et al. [2], twisted spade rudders can reduce the cavitation and improve propulsion efficiency. Ahn et al. [2] reported an X-Twisted rudder which can reduce the rudder cavitation and improve the overall maneuverability.

2.5.4 Concluding Remarks of Rudder Performance

As presented in the previous sections, the rudder performance can be roughly judged in three aspects: ship maneuverability that relates to navigation safety, fuel consumption that affects transport efficiency, and rudder cavitation that determines operation and maintenance cost. The keys to improving the effectiveness of the rudder in ship maneuverability are increasing the rudder inflow velocity or the percentage of the rudder area in the propeller slipstream, enlarging the total rudder area, and improving the rudder hydrodynamic characteristics by changing the profile, the property, and the type. However, the improvement in the ship maneuverability commonly comes at the expense of extra rudder induced resistance. Future research should, therefore, concentrate on the investigation of energy-saving methods of the rudders such as the twisted rudders and the energy-saving bulbs. Considering the cavitation, whole-body profiles and types like the spade rudder with a NACA profile have advantages over the separated ones like the semi-skeg rudder with a flap profile. Even though for common commercial ships, the effectiveness and efficiency of the rudder get higher priority than the cavitation in the design process, a final check of the rudder performance in the cavitation cannot be neglected.

2.6 Summary

Section 2.3 has reviewed the common rudder profiles, namely flat-plate, NACA, HSVA, IFS, fishtail, wedge-tail, and flap. The flat-plate profile is effective with small rudder angles but limited by stiffness issues in the rudder area. Considering the enlargement of inland vessels, the flat-plate profile is not suitable and thus not considered in this monograph. The NACA series has been widely applied and systematically studied. Therefore, it is chosen as the benchmark profile and applied to CFD validation in Chap. 4. The IFS series is taken as a representative of the rudder oriented profiles.

Fishtail and wedge-tail rudders are both designed to have high lift coefficients. Since no publicly available offsets of these profiles were found, the author designed a series of wedge-tail profiles based on the NACA profiles of the same thickness. The flap profile may have superior performance to the classic whole-body profiles but it has disadvantages of hinge moments, mechanical complexity, and maintenance difficulties. The flap profile is a good alternative for ships that require high performance

such as tugs, working boats, and cruise ships. In summary, this monograph chooses
the NACA, the IFS, and the wedge-tail profiles as reference profiles.

Key parameters of a rudder design, more specifically the area, the thickness, the
span, the chord length, and the aspect ratios, should be considered as a whole to
find the most suitable combination. In general, the area should be sufficiently large.
The thickness should be as thin as possible with the prerequisite of satisfying the
structural requirements. The span and chord length should be optimized to ensure
large aspect ratios. This monograph does not vary these detailed design parameters
in each tested rudder configuration but provides an integrated maneuvering model
(Chap. 5) which makes it possible to analyze the impacts of each parameter.

Based on the above-mentioned literature and daily observation of the inland vessels in Chinese and Dutch waterways, it is concluded that the spade type rudder is still
the most suitable for commercial motor vessels because the spade rudders are superior
to other types with regard to the overall performance of hydrodynamics, cavitation,
and maintenance. Considering the enlargement of inland vessels and inland-seagoing
combined vessels, the semi-skeg type is a good alternative that enables a large rudder
area. In this monograph, the impacts of rudder configurations on ship maneuverability are analyzed based on the spade type rudder, but the results are also applicable to
other rudder types.

The interaction between the propeller and the rudder is primarily determined
by their relative positions. Longitudinal and lateral separations affect the incidence
angles and speeds while the vertical operation has little influence. Inland vessels
commonly feature multiple-rudder configurations and the interactions among rudders
should be carefully considered. Impacts of spacing among twin and quadruple rudders
are analyzed in Chap. 4 and their influences on inland vessel maneuverability are
further discussed in Chap. 8.

References

1. Abbott IH, Von Doenhoff AE (1959) Theory of wing sections: including a summary of airfoil
   data, 1st edn. Dover Publications, New York
2. Ahn K, Choi GH, Son DI, Rhee KP (2012) Hydrodynamic characteristics of X-Twisted rudder
   for large container carriers. Int J Nav Arch Ocean Eng 4(3):322–334
3. Alte R, Baur MV (1986) Propulsion Handbuch der Werften (in German). Hansa 18:132
4. Badoe CE, Phillips AB, Turnock SR (2015) Influence of drift angle on the computation of
   hull-propeller-rudder interaction. Ocean Eng 103:64–77
5. Barrass B (2004) Ship design and performance for masters and mates, 1st edn. Elsevier
   Butterworth-Heinemann, Oxford
6. Batten WMJ, Bahaj AS, Molland AF, Chaplin J (2006) Hydrodynamics of marine current
   turbines. Renew Energy 31(2):249–256
7. Batten WMJ, Bahaj AS, Molland AF, Chaplin JR (2008) The prediction of the hydrodynamic
   performance of marine current turbines. Renew Energy 33(5):1085–1096
8. Baudu H (2014) Ship handling, 1st edn. DOKMAR Maritime Publishers BV, Enkhuizen
9. Bertagnolio F, Sørensen NN, Johansen J, Fuglsang P (2001) Wind turbine airfoil catalogue.
   Denmark. Forskningscenter Risoe. Risoe-R; No. 1280 (EN)
10. Bertram V (2012) Practical ship hydrodynamics, 2nd edn. Elsevier, Oxford

11. Bonci M, Viviani M, Broglia R, Dubbioso G (2015) Method for estimating parameters of practical ship manoeuvring models based on the combination of RANSE computations and system identification. Appl Ocean Res 52:274–294

12. Brennen CE (1995) Cavitation and bubble dynamics. Cambridge University Press, Oxford

13. Brix J (1993) Manoeuvring technical manual. Seehafen, Hamburg

14. Broglia R, Dubbioso G, Durante D, Di Mascio A (2013) Simulation of turning circle by CFD: analysis of different propeller models and their effect on manoeuvring prediction. Appl Ocean Res 39:1–10

15. Caldas A, Meis M, Sarasquete A (2011) Numerical analysis of rudder effects upon ducted propeller units. In: 2nd international symposium on marine propulsors. Hamburg, Germany

16. Carlton JS, Radosavljevic D, Whitworth S (2009) Rudder-propeller-hull interaction: the results of some recent research, in-service problems and their solutions. In: 1st symposium on marine propulsors. Trondheim, Norway

17. Champlain JG (1971) Analysis of flapped rudder gap effects. Master thesis, Massachusetts Institute of Technology

18. China Classification Society (2003) Guidelines for inland vessel manoeuvrability (in Chinese). Guidance Notes GD-2003

19. Clarke D, Gedling P, Hine G (1983) Application of manoeuvring criteria in hull design using linear theory. Trans R Inst Nav Arch 125:45–68

20. Cleynen O (2011) Airfoil nomenclature. Accessed 30 Dec 2015, retrieved from https://en.wikipedia.org/wiki/Airfoil

21. Coraddu A, Dubbioso G, Mauro S, Viviani M (2013) Analysis of twin screw ships' asymmetric propeller behaviour by means of free running model tests. Ocean Eng 68:47–64

22. Di Mascio A, Dubbioso G, Notaro C, Viviani M (2011) Investigation of twin-screw naval ships maneuverability behavior. J Ship Res 55(4):221–248

23. Di Mascio A, Dubbioso G, Muscari R, Felli M (2015) CFD analysis of propeller-rudder interaction. In: 25th international ocean and polar engineering conference. Kona, Hawaii, USA, pp 946–950

24. Dubbioso G, Viviani M (2012) Aspects of twin screw ships semi-empirical maneuvering models. Ocean Eng 48:69–80

25. Dubbioso G, Mauro S, Ortolani F (2015) Experimental and numerical investigation of asymmetrical behaviour of rudder/propeller for twin screw ships. In: International conference on marine simulation and ship maneuverability (MARSIM '15). Newcastle upon Tyne, UK

26. Fach K, Bertram V (2007) High-performance simulations for high-performance ships. Ships Offshore Struct 2(2):105–113

27. Felli M, Falchi M (2011) Propeller tip and hub vortex dynamics in the interaction with a rudder. Exp Fluids 51(5):1385–1402

28. Felli M, Roberto C, Guj G (2009) Experimental analysis of the flow field around a propeller-rudder configuration. Exp Fluids 46:147–164

29. Fink MP, Lastinger JL (1961) Aerodynamic characteristics of low-aspect-ratio wings in close proximity to the ground. Technical report, Langley Research Center, Washington, DC, USA

30. Fujii H (1960) Experimental researches on rudder performance (1) (in Japanese). J Zosen Kiokai 107:105–111

31. Fujii H, Tsuda T (1961) Experimental researches on rudder performance (2) (in Japanese). J Zosen Kiokai 109:31–42

32. Gim OS (2013) Assessment of flow characteristics around twin rudder with various gaps using PIV analysis in uniform flow. Ocean Eng 66:1–11

33. Goundar JN, Rafiuddin Ahmed M, Lee YH (2012) Numerical and experimental studies on hydrofoils for marine current turbines. Renew Energy 42:173–179

34. Gregory N, O'Reilly CL (1970) Low-speed aerodynamic characteristics of NACA 0012 aerofoil section, including the effects of uppers-surface roughness simulating hoar frost. Technical report, Ministry of Defence, London, UK

35. Greitsch L (2008) Prognosis of rudder cavitation risk in ship operation. In: 11th numerical towing tank symposium. Brest, France

36. Greitsch L, Eljardt G, Krueger S (2009) Operating conditions aligned ship design and evaluation. In: 1st international symposium on marine propulsors. Trondheim, Norway
37. Hamamoto M, Enomoto T (1997) Maneuvering performance of a ship with VecTwin rudder system. J Soc Nav Arch Jpn, pp 197–204
38. Han JM, Kong DS, Song IH, Lee CS (2001) Analysis of the cavitation flow around the horn-type rudder in the race of a propeller. In: 4th international symposium on cavitation. Pasadena, CA, USA
39. Harris CD (1981) Two-dimensional aerodynamic characteristics of the NACA 0012 airfoil in the langley 8 foot transonic pressure tunnel. Technical report, Langley Research Center, Hampton, Virginia, USA
40. Hasegawa K, Kang D, Sano M, Nabeshima K (2006a) Study on the maneuverability of a large vessel installed with a mariner type Super VecTwin rudder. J Mar Sci Technol 11(2):88–99
41. Hasegawa K, Kang D, Sano M, Nagarajan V, Yamaguchi M (2006b) A study on improving the course-keeping ability of a pure car carrier in windy conditions. J Mar Sci Technol 11:76–87
42. Hasegawa K, Nagarajan V, Kang DH (2006) Performance evaluation of schilling rudder and mariner rudder for pure car carriers (PPC) under wind condition. In: International conference on marine simulation and ship maneuverability (MARSIM '06). Terschelling, The Netherlands, pp 1–10
43. Hochkirch K, Bertram V (2010) Engineering options for more fuel efficient ships. In: 1st international symposium on fishing vessel energy efficiency. Vigo, Spain
44. Hoerner SF (1965) Fluid-dynamic drag: practical information on aerodynamic drag and hydrodynamic resistance, 2nd edn. Hoerner Fluid Dynamics, Bakersfield
45. Hollenbach U, Friesch J (2007) Efficient hull forms: what can be gained? In: 1st International conference on ship efficiency. Hamburg, Germany
46. Hollenbach U, Reinholz O (2011) Hydrodynamic trends in optimizing propulsion. In: 2nd international symposium on marine propulsors. Hamburg, Germany
47. International Maritime Organization (2011) Amendments to the annex of the protocol of 1997 to amend the international convention for the prevention of pollution from ships, 1973, as modified by the protocol of 1978 relating thereto. Resolution MEPC. 203(62), adopted on July 16, 2011
48. International Maritime Organization (2012) SOLAS consolidated edition 2012: consolidated text of the international convention for the safety of life at sea, 1974, and its protocol of 1988: articles, annexes and certificates
49. Jurgens AJ (2005) Static and dynamic effects of rudder-hull-propeller interaction on fast monohulls. In: 8th international conference on fast sea transportation (FAST 2005). St. Petersburg, Russia
50. Kang D, Hasegawa K (2007) Prediction method of hydrodynamic forces acting on the hull of a blunt-body ship in the even keel condition. J Mar Sci Technol 12(1):1–14
51. Kang D, Nagarajan V, Gonno Y, Uematsu Y, Hasegawa K, Shin SC (2011) Installing single-propeller twin-rudder system with less asymmetric maneuvering motions. Ocean Eng 38:1184–1196
52. Kerwin JE, Lewis SD, Oppenheim BW (1974) Experiments on rudders with small flaps in free-stream and behind a propeller. Technical report, Massachusets Institue of Technology, Cambridge, MA, USA
53. Kerwin JE, Mandel P, Lewis SD (1972a) An experimental study of a series of flapped rudders. J Ship Res 16:221–239
54. Kerwin JW, Mandel P, Lewis SD (1972b) Hydrodynamic characteristics of flapped rudders. J Mech Eng Sci 14(7):142–149
55. Khanfir S, Hasegawa K, Lee SK, Jang TS, Lee JH, Cheon SJ (2008) Mathematical model for maneuverability and estimation of hydrodynamic coefficients of twin-propeller twin-rudder ship. In: The Japan society of naval architects and ocean engineers. Osaka, Japan, pp 57–60, 2008K-G4-3
56. Khanfir S, Hasegawa K, Nagarajan V, Shouji K, Lee SK (2011) Manoeuvring characteristics of twin-rudder systems: Rudder-hull interaction effect on the manoeuvrability of twin-rudder ships. J Mar Sci Technol 16:472–490

57. Khanfir S, Nagarajan V, Hasegawa K, Lee SK (2009) Estimation of mathematical model and its coefficients of ship manoeuvrability for a twin-propeller twin-rudder ship. In: International conference on marine simulation and ship maneuverability (MARSIM '09), vol 8. Panama City, Panama, pp 159–166

58. Kijima K, Katsuno T, Nakiri Y, Furukawa Y (1990) On the manoeuvring performance of a ship with the parameter of loading condition. J Soc Nav Arch Jpn 168:141–148

59. Kim YG, Kim SY, Kim HT, Lee SW, Yu BS (2007) Prediction of the maneuverability of a large container ship with twin propellers and twin rudders. J Mar Sci Technol 12(3):130–138

60. Kim HJ, Kim SH, Oh JK, Seo DW (2012) A proposal on standard rudder device design procedure by investigation of rudder design process at major Korean shipyards. J Mar Sci Technol 20(4):450–458

61. Kim JH, Choi JE, Choi BJ, Chung SH (2014) Twisted rudder for reducing fuel-oil consumption. Int J Nav Arch Ocean Eng 6(3):715–722

62. Koç ST, Yılmaz S, Erdem D, Kavsaoğlu MŞ (2011) Experimental investigation of a ducted propeller. In: Proceedings of the 4th European conference for aerospace sciences. Saint Petersburg, Russia

63. Kracht AM (1989) Rudder in the slipstream of a propeller. In: International symposium on ship resistance and powering performance. Shanghai, China, pp 261–270

64. Ladson CL, Hill AS, Sproles D (1996) Computer program to obtain ordinates for NACA airfoils. Technical report, NASA Technical Memorandum 4741, Langley Research Center, Hampton, Virginia, USA

65. Ladson CL (1988) Effects of independent variation of mach and reynolds numbers on the low-speed aerodynamic characteristics of the NACA 0012 airfoil section. Technical report, Langley Research Center, Hampton, Virginia, USA

66. Laurens JM (2003) Unsteady hydrodynamic behaviour of a rudder operating in the propeller slipstream. Ship Technol Res 50(3):141–148

67. Lee SK, Fujino M (2003) Assessment of a mathematical model for the manoeuvring motion of a twin-propeller twin-rudder ship. Int Shipbuild Prog 50(1–2):109–123

68. Lee S, Rhee KP, Choi JW (2011) Design of the roll stabilization controller using fin stabilizers and pod propellers. Appl Ocean Res 33(4):229–239

69. Lee H, Kinnas SA, Gu H, Natarajan S (2003) Numerical modeling of rudder sheet cavitation including propeller/rudder interaction and the effects of a tunnel. In: 5th international symposium on cavitation. Osaka, Japan

70. Lehmann D (2007) Improved propulsion with tuned rudder systems. In: 1st international conference on ship efficiency. Hamburg, Germany

71. Lehmann D (2012) Station keeping with high-performance rudders. In: Dynamic positioning conference 2012. Houston, USA

72. Liu J, Quadvlieg F, Hekkenberg R (2016) Impacts of the rudder profile on manoeuvring performance of ships. Ocean Eng 124:226–240

73. Liu J, Hekkenberg R (2015) Hydrodynamic characteristics of twin-rudders at small attack angles. In: 12th international marine design conference (IMDC). Tokyo, Japan, pp 177–188

74. Liu J, Hekkenberg R, Jiang Z, Chu X (2017) Mesh properties for RANS simulations of aerofoil-shape rudder hydrodynamics. In: 10th international workshop on ship and marine hydrodynamics (IWSH 2017). Keelung, Taiwan

75. Liu J, Quadvlieg F, Hekkenberg R (2015) Impacts of rudder profiles on ship manoeuvrability. In: International conference on marine simulation and ship maneuverability (MARSIM '15). Newcastle upon Tyne, UK

76. Loftin LK, Smith HA (1949) Aerodynamic characteristics of 15 NACA airfoil sections at seven reynolds numbers from 0.7×10^6 to 9.0×10^6. Technical report, National Advisory Committee for Aeronautics, Washington, DC, USA

77. Lübke L (2009) Investigation of a semi-balanced rudder. Ship Technol Res 56(2):69–86

78. Mandel P (1967) Ship maneuvering and control. Ph.D thesis, Stevens Institute of Technology

79. Mccroskey WJ (1987) A critical assessment of wind tunnel results for the NACA 0012 airfoil. Technical report, National Aeronautics and Space Administration, Moffett Field, CA, USA

80. Mewis F, Klug H (2004) The challenge of very large container ships: a hydrodynamic view. In: Keil H, Lehmann E (eds) 9th symposium on practical design of ships and other floating structures. Luebeck Travemuende, Germany, pp 173–181

81. Minchev A, Schmidt M, Schnack S (2013) Contemporary bulk carrier design to meet IMO EEDI requirements. In: Binns J, Brown R, Bose N (eds) 3rd international symposium on marine propulsors. Launceston, Tasmania, pp 283–291

82. Molland AF (1977) The free-stream characteristics of a semi-balanced ship skeg-rudder. Technical report, University of Southampton, Southampton, UK

83. Molland AF (1978) Further free-stream characteristics of semi-balanced ship skeg-rudders. Technical report, University of Southampton, Southampton, UK

84. Molland AF, Turnock SR (1991) Wind tunnel investigation of the influence of propeller loading on ship rudder performance. Technical report, University of Southampton, Southampton, UK

85. Molland AF, Turnock SR (1992) Further wind tunnel investigation of the influence of propeller loading on ship rudder performance. Technical report, University of Southampton, Southampton, UK

86. Molland AF, Turnock SR (1993) Wind tunnel tests on the influence of propeller loading on ship rudder performance: four quadrant operation. Low and zero speed operation. Technical report, University of Southampton, Southampton, UK

87. Molland AF, Turnock SR (1993) Wind tunnel investigation of the influence of propeller loading on a semi-balanced skeg rudder. Technical report, University of Southampton, Southampton, UK

88. Molland AF, Turnock SR (1995) Wind tunnel tests on the effect of a ship hull on rudder-propeller performance at different angles of drift. Technical report, University of Southampton, Southampton, UK

89. Molland AF, Turnock SR (1995) Some effects of rudder-propeller-hull arrangements on manoeuvring and propulsion. In: Oosterveld MWC, Tan SG (eds) 6th international symposium on practical design of ships and mobile units, vol 1. Seoul, Korea, pp 333–345

90. Molland AF, Turnock SR, Smithwick JET (1995) Wind tunnel tests on the influence of propeller loading and the effect of a ship hull on skeg-rudder performance. Technical report, University of Southampton, Southampton, UK

91. Molland AF, Turnock SR (2002) Flow straightening effects on a ship rudder due to upstream propeller and hull. Int Shipbuild Prog 49(3):195–214

92. Molland AF, Turnock SR (2007) Marine rudders and control surfaces: principles, data, design and applications, 1st edn. Elsevier Butterworth-Heinemann, Oxford

93. Molland AF, Bahaj AS, Chaplin JR, Batten WMJ (2004) Measurements and predictions of forces, pressures and cavitation on 2-D sections suitable for marine current turbines. Proc Inst Mech Eng, Part M: J Eng MaritE Environ 218(2):127–138

94. Motoki A, Ohashi K, Nobuaki S (2015) Effects of rudder horn and propeller hub vortex for CFD manoeuvring simulations. In: International conference on marine simulation and ship maneuverability (MARSIM '15). Newcastle upon Tyne, UK

95. Nagarajan V, Kang DH, Hasegawa K, Nabeshima K (2008) Comparison of the mariner Schilling rudder and the mariner rudder for VLCCs in strong winds. J Mar Sci Technol 13:24–39

96. Nagarajan V, Kang DH, Hasegawa K, Nabeshima K, Arii T (2009) A proposal for propulsion performance prediction of a single-propeller twin-rudder ship. J Mar Sci Technol 14:296–309

97. Natarajan S (2003) Computational modeling of rudder cavitation and propeller/rudder interaction. Master thesis, University of Texas

98. Nienhuis U (1987) Passieve Manoeuvreerhulpmiddelen: Open Water Proeven met Roer (in Dutch). Technical report, Maritime Research Institute Netherlands (MARIN), Wageningen, The Netherlands

99. Olson CR (1955) Effects of various linkage ratios on the free-stream hydrodynamic characteristics of an all-movable flapped rudder. Technical report, David Taylor Model Basin, Washington, DC, USA

100. Oppenheim BW (1974) A theoretical and experimental investigation of the performance of flapped rudders. Master thesis, Massachusetts Institute of Technology
101. Pecoraro A, Di Felice F, Felli M, Salvatore F, Viviani M (2015) An improved wake description by higher order velocity statistical moments for single screw vessel. Ocean Eng 108:181–190
102. Pelletier A, Mueller TJ (2000) Low Reynolds number aerodynamics of low-aspect-ratio, thin/flat/cambered-plate wings. J Aircr 37(5):825–832
103. Perez T (2005) Ship motion control: course keeping and roll stabilisation using rudder and fins. Springer, London
104. Phillips AB, Turnock SR, Furlong M (2010) Accurate capture of rudder-propeller interaction using a coupled blade element momentum-RANS approach. Ship Technol Res 57(2):128–139
105. Quadvlieg F (2013) Mathematical models for the prediction of manoeuvres of inland ships: does the ship fit in the river? In: Rigo P, Wolters M (eds) Smart rivers 2013. PIANC, Liège, Belgium/Maastricht, The Netherlands, pp 1871–1879
106. Ram BRR, Surendran S, Lee SK (2015) Computer and experimental simulations on the fin effect on ship resistance. Ships Offshore Struct 10(2):121–131
107. Reichel M (2009) Influence of rudder location on propulsive characteristics of a single screw container ship. In: Koushan K, Steen S (eds) 1st international symposium on marine propulsors. Norwegian Marine Technology Research Institute (MARINTEK), Trondheim, Norway, pp 1–6
108. Ren R, Zou Z, Wang X (2014) A two-time scale control law based on singular perturbations used in rudder roll stabilization of ships. Ocean Eng 88:488–498
109. Rhee SH, Lee C, Lee HB, Oh J (2010) Rudder gap cavitation: fundamental understanding and its suppression devices. Int J Heat Fluid Flow 31(4):640–650
110. Sarasquete A, Collazo AC, Coache S, Meis M, Ruiz V (2012) Increased energy efficiency of the fishing fleet due to improved hydrodynamic performance. In: 2nd international symposium on fishing vessel energy efficiency. Vigo, Spain
111. Sathaye SS (2004) Lift distributions on low aspect ratio wings at low Reynolds numbers. Master thesis, Worcester Polytechnic Institute, Worcester, MA, USA
112. Schilling K (1963) Rudder control arrangement. U S Pat Off 3(101):693
113. Schilling K, Rathert H (1978) Dual rudder assembly. U S Pat Off 4(085):694
114. Schneekluth H, Bertram V (1998) Ship design for efficiency and economy, 2nd edn. Elsevier Butterworth-Heinemann, Oxford
115. Sharif MT, Roberts GN, Sutton R (1995) Sea-trial experimental results of fin/rudder roll stabilisation. Control Eng Pract 3(5):703–708
116. Sharif MT, Roberts GN, Sutton R (1996) Final experimental results of full scale fin/rudder roll stabilisation sea trials. Control Eng Pract 4(3):377–384
117. Shen YT, Jiang CW, Remmers KD (1997) A twisted rudder for reduced cavitation. J Ship Res 41(4):260–272
118. Shen YT, Remmers KD, Jiang CW (1997) Effects of ship hull and propeller on rudder cavitation. J Ship Res 41(3):172–180
119. Simonsen CD (2000) Rudder, propeller and hull interaction by RANS. Ph.D thesis, Technical University of Denmark
120. Söding H (1982) Prediction of ship steering capabilities. Schiffstechnik 29(1):3–29
121. Stuck A, Turnock S, Bressloff N (2004) An evaluation of the RANS method for the prediction of steady ship rudder performance compared to wind tunnel measurements. Technical report, University of Southampton, Southampton, UK
122. Surendran S, Kiran V (2006) Technical note Studies on the feasibilities of control of ship roll using fins. Ships Offshore Struct 1(4):357–365
123. Surendran S, Kiran V (2007) Control of ship roll motion by active fins using fuzzy logic. Ships Offshore Struct 2(1):11–20
124. Tabaczek T (2010) Numerical simulation of planar motion of a twin-screw inland waterway vessel in shallow water. In: 18th international conference on hydrodynamics in ship design. Safety and Operation, Gdansk, Poland, pp 37–50

125. Takekoshi Y, Kawamura T, Yamaguchi H, Maeda M, Ishii N, Kimura K, Taketani T, Fujii A (2005) Study on the design of propeller blade sections using the optimization algorithm. J Mar Sci Technol 10(2):70–81
126. Thieme H (1965) Design of Ship Rudders (Zur Formgebung von Shiffsrudern). Technical report, Shipbuilding Institute, University of Hamburg, Washington, DC, USA
127. Thieme H (1962) Zur formgebung von shiffsrudern. Jahrbuch der Schiffbautechnische Gesellschaft 56:381–426
128. Timmer WA (2010) Aerodynamic characteristics of wind turbine blade airfoils at high angles-of-attack. In: 3rd EWEA conference-torque 2010: the science of making torque from wind. Heraklion, Crete, Greece, pp 71–78
129. Torres GE, Mueller TJ (2004) Low-aspect-ratio wing aerodynamics at low Reynolds numbers. AIAA J 42(5):865–878
130. Turnock SR (1990) A test rig for the investigation of ship propeller/rudder interactions. Technical report, University of Southampton, Southampton, UK
131. Turnock SR (1993) Prediction of ship rudder-propeller interaction using parallel computations and wind tunnel measurements. Ph.D thesis, University of Southampton
132. Turnock SR, Wright AM (2000) Directly coupled fluid structural model of a ship rudder behind a propeller. Mar Struct 13(1):53–72
133. Ueno M, Tsukada Y (2015) Rudder effectiveness and speed correction for scale model ship testing. Ocean Eng 109:495–506
134. Ueno M, Tsukada Y, Kitagawa Y (2014) Rudder effectiveness correction for scale model ship testing. Ocean Eng 92:267–284
135. Van Amerongen J (1991) Ship rudder roll stabilization. In: Papageorgiou M (ed) Concise encyclopedia of traffic and transportation systems. Pergamon, Amsterdam, pp 448–454
136. Van Amerongen J, Van Der Klugt PGM, Van Nauta Lemke HR (1990) Rudder roll stabilization for ships. Automatica 26(4):679–690
137. Van Beek T (2004) Technology guidelines for efficient design and operation of ship propulsors. Marine News, Wärtsillä Propulsion, Netherlands BV 1:14–19
138. Van Nguyen T, Ikeda Y (2014) Development of fishtail rudder sections with higher maximum lift coefficients. In: 24th international ocean and polar engineering conference. Busan, Korea, pp 940–947
139. Van Nguyen T, Ikeda Y (2013) Hydrodynamic characteristic of rudder sections with high lift force. J Jpn Soc Nav Arch Ocean Eng, pp 403–406
140. Van Nguyen T, Ikeda Y (2014) Hydrodynamic characteristics of rudder sections with high lift force, part 2: the wedge tail shapes. J Jpn Soc Nav Arch Ocean Eng, pp 171–174
141. Veritas DN (2000) Hull equipment and appendages: stern frames, rudders and steering gears. Rules for Classification of Steel Ships, pp 6–28, part 3, chapter 3, section 2
142. Whicker LF, Fehlner LF (1958) Free-stream characteristics of a family of low-aspect-ratio. All-movable control surfaces for application to ship design. Technical report, David Taylor Model Basin, Washington, DC, USA
143. Yang H, Lee J, Kim K (2015) Numerical and Experimental study on the rudder force of a twisted rudder. In: international conference on marine simulation and ship maneuverability (MARSIM '15). Newcastle upon Tyne, UK
144. Yasukawa H, Yoshimura Y (2015) Introduction of MMG standard method for ship maneuvering predictions. J Mar Sci Technol 20(1):37–52. https://doi.org/10.1007/s00773-014-0293-y
145. Yılmaz S, Erdem D, Kavsaoğlu MŞ (2013) Effects of duct shape on a ducted propeller performance. In: 51st AIAA aerospace sciences meeting including the new horizons forum and aerospace exposition. Grapevine, Texas, USA, pp 1–11
146. Yoo WJ, Yoo BY, Rhee KP (2006) An experimental study on the maneuvering characteristics of a twin propeller/twin rudder ship during berthing and unberthing. Ships Offshore Struct 1(3):191–198
147. Yoshimura Y, Sakurai H (1989) Mathematical model for the manoeuvring ship motion in shallow water (3rd report): manoeuvrability of a twin-propeller twin-rudder ship. J Kansai Soc Nav Arch, Jpn 211:115–126

Chapter 3
Computational Fluid Dynamics Methods of Ship Rudders

Inland vessels have more complex rudder configurations than seagoing ships. Currently, the rudder profile, the number of rudders, and the spacing between rudders are not considered in the empirical methods for rudder force calculation. To analyze the impacts of the rudder configurations on inland vessel maneuverability, the hydrodynamic characteristics, for instance, lift and drag coefficients, of each configuration need to be calculated specifically. Conventional methods to study rudder hydrodynamics are wind-tunnel and open water tests. Due to the high cost and the scale effects of the model tests, the Computational Fluid Dynamics (CFD) method is considered as an increasingly attractive alternative. CFD methods are roughly classed in a hierarchy of lifting-line, boundary element, Reynolds-Averaged Navier–Stokes (RANS), Large Eddy Simulation (LES), and Direct Numerical Simulation (DNS) methods.

The lifting-line method and similar potential flow-based approaches describe the velocity field as the gradient of a scalar potential function. Assuming the flow is irrotational, the flow vorticity is expressed as a Laplace equation. With these assumptions, the lifting-line method has the advantages of implementation and computational time. However, the application of lifting-line methods can only be applied to high aspect ratios while ship rudders typically have a geometric aspect ratio in the range of 1–3. Due to the assumption of potential flow, the lifting-line methods and boundary element methods cannot model the viscous flow effects, such as frictional drag, flow separation, and stall [1, p. 237]. The LES method simulates the large-scale eddies individually to reduce the length scale ranges of the solution, reducing the cost of computation while the DNS method presents all the flow motion. However, both LES and DNS methods require very fine mesh and small-time steps, making them expensive in engineering applications.

This chapter gives a general description of CFD methods and provides a step-by-step procedure of the RANS method applied to rudder hydrodynamic analysis. The

© Springer Nature Switzerland AG 2020
J. Liu, *Mathematical Modeling of Inland Vessel Maneuverability Considering Rudder Hydrodynamics*, https://doi.org/10.1007/978-3-030-47475-1_3

presented method uses a time-averaged Reynolds decomposition, which assumes that all the components of the flow velocity and pressure consist of a mean value and a bounded fluctuation to represent turbulence. Compared to the above-mentioned CFD methods, the RANS method cannot only analyze the induced lift and drag of a rudder with complex geometry, but also model the flow separation, and the stall angle with much fewer requirements of calculation resources than the LES and DNS methods. To obtain reliable results, the RANS simulations should be carefully configured in the turbulence modeling (Sect. 3.2), the boundary conditions (Sect. 3.3), the numerical solver (Sect. 3.4), and the mesh generation (Sect. 3.5).

3.1 Governing Equations of Fluid Dynamics

In common marine applications of CFD, global pressures and velocity components are calculated, based on which the forces and moments acting on the vehicle or the structure of interest can be resolved [2]. According to MARNET-CFD [2], Versteeg and Malalasekera [3], the conservation laws of physics and the mathematical statements for a three-dimensional, compressible, unsteady, viscous, turbulent, continuum fluid with heat transfer are as follows:

- **Mass conservation of a fluid:**
  The mass of the fluid flow is conserved, which says the rate at which mass enters a system equals the rate at which mass leaves the system. To represent the conservation of mass, the continuity equation is expressed as the following:

$$\frac{\partial \rho}{\partial t} + \nabla dot (\rho U) = 0, \tag{3.1}$$

  where ρ is the fluid density, t is time, ∇ stands for $\left(\frac{\partial}{\partial x}, \frac{\partial}{\partial y}, \frac{\partial}{\partial z} \right)$, and $U = (u, v, w)$ is the fluid speed.
- **Momentum conservation of a fluid**:
  The rate of change of momentum equals the sum of the forces on a fluid particle, i.e. Newton's second law. To represent the conservation of momentum in x, y, and z directions, the momentum equations are revealed as follows:

$$\left. \begin{aligned} \frac{\partial (\rho u)}{\partial t} + \nabla \cdot (\rho u U) &= -\frac{\partial p}{\partial x} + \frac{\partial \tau_{xx}}{\partial x} + \frac{\partial \tau_{yx}}{\partial y} + \frac{\partial \tau_{zx}}{\partial z} + \rho f_x \\ \frac{\partial (\rho v)}{\partial t} + \nabla \cdot (\rho v U) &= -\frac{\partial p}{\partial y} + \frac{\partial \tau_{xy}}{\partial x} + \frac{\partial \tau_{yy}}{\partial y} + \frac{\partial \tau_{zy}}{\partial y} + \rho f_y \\ \frac{\partial (\rho w)}{\partial t} + \nabla \cdot (\rho w U) &= -\frac{\partial p}{\partial z} + \frac{\partial \tau_{xz}}{\partial x} + \frac{\partial \tau_{yz}}{\partial y} + \frac{\partial \tau_{zz}}{\partial z} + \rho f_z, \end{aligned} \right\} \tag{3.2}$$

where p is the pressure, τ with subscripts x, y, and z are the viscous stresses, $f = (f_x, f_y, f_z)$ is body force.

- **Energy conservation of a fluid:**

The rate of change of energy equals the sum of the rate of heat added to and the rate of work done on a fluid particle, i.e. first law of thermodynamics. To present the conservation of energy, the energy equations are as follows:

$$
\begin{aligned}
\frac{\partial}{\partial t}\left(\rho\left(i+\frac{U^2}{2}\right)\right)+\nabla\cdot\left(\rho U\left(i+\frac{U^2}{2}\right)\right) &= \rho\dot{q}+\rho f U \\
&-\frac{\partial(up)}{\partial x}-\frac{\partial(vp)}{\partial y}-\frac{\partial(wp)}{\partial z} \\
&+\frac{\partial}{\partial x}\left(k\frac{\partial T}{\partial x}\right)+\frac{\partial}{\partial y}\left(k\frac{\partial T}{\partial y}\right)+\frac{\partial}{\partial z}\left(k\frac{\partial T}{\partial z}\right) \\
&+\frac{\partial(u\tau_{xx})}{\partial x}+\frac{\partial(u\tau_{yx})}{\partial y}+\frac{\partial(u\tau_{zx})}{\partial z} \\
&+\frac{\partial(v\tau_{xy})}{\partial x}+\frac{\partial(v\tau_{yy})}{\partial y}+\frac{\partial(v\tau_{zy})}{\partial z} \\
&+\frac{\partial(w\tau_{xz})}{\partial x}+\frac{\partial(w\tau_{yz})}{\partial y}+\frac{\partial(w\tau_{zz})}{\partial z},
\end{aligned}
\tag{3.3}
$$

where i is the internal energy per unit mass, $\dot{q}$ is the rate of volumetric heat addition per unit mass, k is the thermal conductivity, and T is the temperature.

- **Motion dynamics of a fluid:**

The motion of a three-dimensional fluid is described by the above-mentioned 5 partial differential equations in 7 unknowns, which are u, v, w, p, T, ρ, and i. Two additional algebra equations of state are needed to link the 4 thermodynamic variables p, T, ρ, and i as follows:

$$
\left.\begin{aligned}
p &= \rho(\rho, T) \\
i &= i(\rho, T).
\end{aligned}\right\}
\tag{3.4}
$$

When proper initial and boundary conditions are supplied, the above governing equations can be solved. Furthermore, the governing equations can be simplified according to the nature of the problem and the properties of the fluid. Commonly for study on marine hydrodynamics, water is taken as the working fluid and regarded as an incompressible and isothermal Newtonian fluid, i.e. ρ and T are constants. With these assumptions, only conservation of mass and momentum are necessary to resolve as follows:

- **Continuity equations of water:**

$$
\nabla \cdot U = 0.
\tag{3.5}
$$

- **Momentum equations of water:**

$$
\left.
\begin{aligned}
\rho \frac{Du}{Dt} &= -\frac{\partial p}{\partial x} + \mu \nabla^2 u + \rho f_x \\
\rho \frac{Dv}{Dt} &= -\frac{\partial p}{\partial y} + \mu \nabla^2 v + \rho f_y \\
\rho \frac{Dw}{Dt} &= -\frac{\partial p}{\partial z} + \mu \nabla^2 w + \rho f_z,
\end{aligned}
\right\}
\tag{3.6}
$$

where $\frac{D}{Dt}$ is used for simplicity as the following:

$$
\frac{D}{Dt} = \frac{\partial}{\partial t} + u\frac{\partial}{\partial x} + v\frac{\partial}{\partial y} + w\frac{\partial}{\partial z}.
\tag{3.7}
$$

However, the above equations are on one side subject to numerical instability and on the other side not suitable for the direct computation of turbulent flows. In practice, it is common to assume that the flow velocity (u, v, and w) and the pressure (p) consist of a mean component (U, V, W, and P) and a unsteady component (u', v', w', and p'), which is the well-known Reynolds-Averaged Navier–Stokes (RANS) method, as the following:

$$
u = U + u' \quad v = V + v' \quad w = W + w' \quad p = P + p'.
\tag{3.8}
$$

Then the time-averaged format of the governing equations are given as:

- **RANS continuity equations for water:**

$$
\frac{\partial U}{\partial x} + \frac{\partial V}{\partial y} + \frac{\partial W}{\partial z} = 0.
\tag{3.9}
$$

- **RANS momentum equations for water:**

$$
\left.
\begin{aligned}
\rho \frac{DU}{Dt} &= -\frac{\partial P}{\partial x} + \mu \nabla^2 U - \rho \left(\frac{\partial \overline{u'^2}}{\partial x} + \frac{\partial \overline{u'v'}}{\partial y} + \frac{\partial \overline{u'w'}}{\partial z} \right) \\
\rho \frac{DV}{Dt} &= -\frac{\partial P}{\partial y} + \mu \nabla^2 V - \rho \left(\frac{\partial \overline{u'v'}}{\partial x} + \frac{\partial \overline{v'^2}}{\partial y} + \frac{\partial \overline{v'w'}}{\partial z} \right) \\
\rho \frac{DW}{Dt} &= -\frac{\partial P}{\partial z} + \mu \nabla^2 W - \rho \left(\frac{\partial \overline{u'w'}}{\partial x} + \frac{\partial \overline{v'w'}}{\partial y} + \frac{\partial \overline{w'^2}}{\partial z} \right),
\end{aligned}
\right\}
\tag{3.10}
$$

where $\overline{u'^2}$, $\overline{v'^2}$, $\overline{w'^2}$, $\overline{u'v'}$, $\overline{u'w'}$, and $\overline{v'w'}$ are additional Reynolds stresses acting on both normal and shear directions to represent the nature of the turbulence.

Thus far, the theory behind the RANS method utilized in this monograph is introduced. As the working fluid, water is assumed to be incompressible and isothermal

Newtonian fluid, the complete governing equations are simplified to governing equations of incompressible water. In practice, the RANS method is widely applied, which assumes the unsteady value as a combination of a mean value and a fluctuation. Then, the governing equations are expressed in a time-averaged format and the question becomes how to express the six Reynolds stress terms in known values. The applied procedure, i.e. turbulence modeling, is discussed in the next section.

3.2 Turbulence Modeling

Due to the complexity and vast computer resources needed for solving full unsteady governing equations, the Reynolds-Averaged Navier–Stokes (RANS) method is widely applied to capture the essence of physics with the minimum amount of complexity. Turbulence models are developed to express the mean effects of turbulent stresses. The turbulent characteristics of the flow play a crucial role in the determination of the fractional drag, the flow separation, the laminar to turbulent transition, and the thickness of boundary layers. No single turbulence model can be universally applied to any turbulent flow simulations [1, p. 248].

To utilize the RANS method for studies of rudder hydrodynamics, the turbulence model has to be chosen according to its suitability to the flow properties, such as the Reynolds number and the angle of attack as discussed in Sect. 2.2. The transition from laminar to turbulent flow mainly depends on the Reynolds number, the body roughness, and the turbulence existing in the income flow. The transition threshold of the Reynolds number is about 0.5×10^6, which is much smaller than the practical range of ship rudders.

Owing to the shallow water, currents, ship making waves, and roughness of the rudder body, rudders are mainly operated in highly turbulent flows. Furthermore, the high turbulence model also poses requirements on the applied mesh, which in turn affects the accuracy and the cost of the simulations. In engineering applications, the turbulence model is commonly chosen from Spalart–Allmaras (SA), Standard k-ε, RNG k-ε, Realizable k-ε, Standard k-ω, and k-ω SST models. Table 3.1 summaries the applied and compared turbulence models in literature. The theory behind these models is briefly introduced in the following sections.

3.2.1 Spalart–Allmaras

The Spalart–Allmaras (SA) model is a one-equation model that solves the transport equation for the turbulent kinematic viscosity [22]. Compared with the Standard k-ε and RNG k-ε models, Stuck et al. [4, p. 22] confirmed that, for large angles of attack, only the SA model can predict the flow detachment while the other two models can at most show the flow stagnates without detaching. The SA model is originally

Table 3.1 Applied or compared turbulence models for RANS simulations in literature

Year	Literature	Applied or compared models
2004	Stuck et al. [4]	SA, Standard k-ε, and RNG k-ε
2010	Wasberg and Reif [5]	SA
2011	Krasilnikov et al. [6]	k-ω SST
2011	Castro et al. [7]	k-ω SST
2012	Eleni et al. [8]	SA, Realizable k-ε, and k-ω SST
2012	Karim and Ahmmed [9]	RNG k-ε
2013	Broglia et al. [10]	SA
2013	Van Nguyen and Ikeda [11]	k-ω SST
2014	Van Nguyen and Ikeda [12]	k-ω SST
2014	Van Nguyen and Ikeda [13]	k-ω SST
2014	Shenoi et al. [14]	k-ω SST
2014	Wang and Zou [15]	RNG k-ε
2015	Van Nguyen and Ikeda [16]	Standard k-ε
2015	Tezdogan et al. [17]	Standard k-ε
2015	Badoe et al. [18]	k-ω SST
2015	Broglia et al. [19]	SA
2016	Dubbioso et al. [20]	SA
2016	Bhattacharyya et al. [21]	k-ω SST

designed for wall-bounded aerodynamic flows. The original SA model is effective for low-Reynolds-number applications, requiring the viscous affected region of the boundary layer to be properly resolved [8].

3.2.2 Standard k-ε, RNG k-ε, and Realizable k-ε

The Standard k-ε model was first given by Launder and Spalding [23]. It is a two-equation model that includes two additional transported variables k and ε, where k is the turbulent kinetic energy per mass and ε is the rate of dissipation of k, to represent the turbulent properties of the flow. In addition, ε specifies the scale of the turbulence and k determines the energy in the turbulence. The Standard k-ε model is widely used in industry flow and heat transfer simulations owing to its robustness, economy, and accuracy for a wide range of turbulent flows [8]. In addition, Quérard et al. [24] noted that the Standard k-ε model can save nearly 25 % CPU time compared to the Standard k-ω model.

Date [25] concluded the weaknesses of applying the Standard k-ε model to rudder applications are the over-predicted turbulent kinetic energy and the under-predicted flow separation, particularly concerning flow stagnation, stall, and reattachment phenomena. MARNET-CFD [2] indicated that the Standard k-ε is generally applicable only to high-Reynolds-number flows and suggested to only use the Standard k-ε for preliminary assessments of steady ship flows. Molland and Turnock [1, p. 249] further pointed that the Standard k-ε is not able to capture laminar and transitional flows, which is an important drawback in the prediction of the rudders that are often operating at a transitional Reynolds number.

To overcome the shortcomings of the Standard k-ε model, Yakhot et al. [26] introduced the RNG k-ε model and Shih et al. [27] developed the Realizable k-ε model. The RNG model has a modified equation of the dissipation rate to account for different scales of motion while the Realizable k-ε model consists of a new model dissipation rate equation and a new realizable eddy viscosity formulation. The Realizable k-ε is more widely applied than the RNG k-ε model in engineering applications. Through the review of the applied turbulence models in Table 3.1, it is noted that the Standard k-ε model and its variations commonly require less computation time and resources than the k-ω SST model, especially for 3D simulations.

3.2.3 Standard k-ω and k-ω SST

The standard k-ω model is based on the Wilcox [28, 29] methods. It is a two-equation model that uses k and ω to predict turbulence, where k is as the same as that for k-ε models while ω is the specific rate of dissipation of the turbulence kinetic energy (k) into internal thermal energy. The standard k-ω model incorporates low-Reynolds-number effects, compressibility, and shear flow spreading [8]. Menter [30] developed the k-ω SST model which effectively combines the strengths of the standard k-ω in the near-wall region and the Standard k-ε in the far-field. Therefore, the k-ω SST model can achieve more accurate and reliable results than the standard k-ω model. For simulations of the NACA 0012 profile, Eleni et al. [8] indicated that the k-ω SST model is more appropriate than the SA model and the Realizable k-ε model.

3.2.4 Concluding Remarks of Turbulence Models

As a summary, the k-ω SST model is applied for 2D simulations in Sect. 4.2 because it gives better results than other models with affordable computation requirements. The Realizable k-ε model is used for 3D simulations in Sect. 4.3 owing to its superiority in computation time and resources to other models, which is only especially important for 3D simulations that have a large number of cells. The SA model is effective for low-Reynolds-number applications and thus not suitable for the high-Reynolds-number simulations of rudders. Standard k-ε, RNG k-ε, and Standard k-ω models are not considered as they are gradually superseded by the k-ω SST and Realizable k-ε models.

3.3 Boundary Conditions

Boundary conditions are needed to define the initial and boundary states of variables for the governing equations and the turbulence models. Correct selections of the boundary conditions are crucial to the physical implementation, the calculation accuracy, and the convergence time. The following paragraphs introduce the applied boundary conditions in this monograph, specifically velocity inlet, pressure outlet, wall, and symmetry.

3.3.1 Velocity Inlet

The velocity inlet specifies the direction and the value of the income flow to the rudder. For open water simulations of rudders, the inlet velocity can be set according to the test Reynolds number. In this monograph, a Reynolds number of 6×10^6 is taken as the threshold value above which the rudder hydrodynamics are not significantly affected as discussed in Sect. 2.2. Otherwise, the inlet velocity is calculated according to the test Reynolds number specified in the validation data.

3.3.2 Pressure Outlet

The pressure outlet condition defines the static pressure at the outlet boundary, which is the same as the static pressure of the environment into which the flow exhausts. The static pressure is assumed to be constant at the outlet and atmospheric. Furthermore, the position of the outlet should be sufficiently far away from the region of interest to avoid back flow, which affects the accuracy of the results.

3.3.3 Wall

The wall condition defines that the tangential velocity of the fluid equals the wall velocity and the normal velocity of the fluid is zero. Considering the viscous effects, the near-wall flow is roughly divided into a laminar layer, a buffer layer, and a turbulent layer. The non-dimensional wall distance (y^+) is commonly used to express the fineness of the mesh for the boundary layers. Different turbulence models have different suitable ranges of y^+. Thus, the value of y^+ has to be carefully considered in the mesh generation.

A rule of thumb is to set y^+ around 1 or in the range of 30–200, considering the structure of the turbulent flow. To obtain the desired y^+, the initial cell height (y_i), which is the distance from the wall to the first mesh point, needs to be determined. For Re smaller than 1×10^9, y_i can be estimated as the following:

$$y_i = \frac{\sqrt{2}C_R y^+}{Re}(2 \log_{10} Re - 0.65)^{1.15}, \tag{3.11}$$

where C_R is the rudder chord length for open water studies of the rudder hydrodynamics. In this monograph, y^+ for the k-ω SST model is smaller than 1 and y^+ for the Realizable k-ε model is around 35.

3.3.4 Symmetry

The symmetry condition assumes that the same physical processes exist on both sides of the boundary. No flow or scalar flux can cross the symmetry boundary. If the viscous effect is out of interest, the symmetry condition can be applied as an alternative to the wall condition to reduce computational effort. In that case, the symmetry condition is a no shear wall condition.

3.3.5 Concluding Remarks of Boundary Conditions

The applied boundary conditions for 2D and 3D simulations are shown in Fig. 3.1. Velocity inlet configures the velocity of the uniform inflow for open water tests. The pressure outlet is set as the same as the static pressure of the environment. The wall boundary specifies the geometry of the rudder and accounts for the viscous effects of the rudder and the bottom of the domain in the validation study of the 3D RANS method. The symmetry boundary (no shear wall) is used for the side walls.

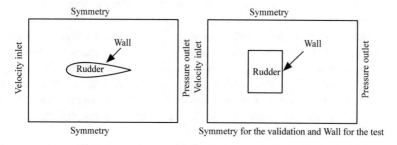

Fig. 3.1 Applied boundary conditions in 2D (left) and 3D (right) simulations

3.4 Numerical Solvers

To represent the conservation principles, differential equations are needed in a discretization form on given mesh. A generic transport equation is commonly applied to arbitrary dependent variable ϕ in form of the transient term, the convection term, the diffusion term, and the source term as the following:

$$\underbrace{\frac{\partial \rho \phi}{\partial t}}_{\text{Transient term}} + \underbrace{\nabla \cdot (\rho \mathbf{u} \phi)}_{\text{Convention term}} = \underbrace{\nabla \cdot (\Gamma \nabla \phi)}_{\text{Diffusion term}} + \underbrace{S_\phi}_{\text{Source term}}, \tag{3.12}$$

where ρ is the water density, $\mathbf{u}$ is the velocity vector, Γ is the diffusion coefficient for ϕ, and S_ϕ is the source or sink of ϕ per unit volume.

To represent the conservation principles, differential equations are needed in a discretization form on given meshes. Discretization techniques, such as the finite difference method, the finite element method, and the finite volume method, are applied to transform each item in Eq. 3.12 to a set of algebraic equations for numerical solvers. In this monograph, the finite volume method is chosen as it is by far the most common approach in current CFD codes [2]. Details of the finite volume method are presented by Versteeg and Malalasekera [3].

Both pressure-based and density-based solvers are both commonly implemented in the commercial package. The density-based solver has the advantage of presenting the shock effects for high-speed compressible fluid over the pressure-based solver. For incompressible viscous water, the pressure-based solver is recommended. The pressure-based solver is separated into segregated and coupled algorithms. The segregated scheme solves the governing equations sequentially while the coupled scheme solves them in a coupled way.

Since the segregated algorithm stores and solves the equations of each variable in the memory one at a time, it is more memory efficient than the coupled solver which has to put all the relevant equations in the memory at the same time. However, the advantage of memory efficiency is obtained at the expense of slow convergence. Kelecy [31] shows that the coupled solver requires more memory, resulting in a long time for each iteration than the segregated one, but the coupled solver may achieve convergence with much fewer iterations than the segregated solver. Additionally, as the interest of the presented simulations is in the performance of the system, namely the hydrodynamics characteristics of rudders, rather than specific changing behaviors, for example, vortex shedding and turbulent studies, steady simulations are favorable than unsteady ones, considering the cost of computational time and resources. All in all, this monograph uses a pressure-based coupled steady solver with the finite volume method.

Discretization schemes are used to discretize governing equations on the discrete form of grid cells, on which solvers are applied to obtain numerical solutions. The finite volume methods (VLM) subdivides the computational domain into several finite volumes or cells. Within each cell, the numerical implementations are applied to conserve mass and momentum. Unlike the potential flow methods, which construct functions on the bounding surfaces, the RANS equations are solved at the cell nodes or the center of each volume in VLM methods. Thus, an approximation is needed to estimate the values on the boundaries, which mainly depend on the discretization schemes and the mesh orthogonality. Considering the impacts of discrete order on solutions, a trade-off has to be made between scheme robustness and solution accuracy.

Commonly, low order schemes are more robust than high order ones, which helps in convergence. Especially for complex geometries, in which case a fine mesh is hard to achieve, low order schemes have fewer convergence problems for poor mesh quality than higher orders. However, lower schemes may suffer more from numerical diffusion with low accuracy. A first-order upwind scheme may lead to stronger numerical diffusion fluxes, while a higher-order scheme may encounter larger dispersion errors.

Higher-order schemes need fewer cells and provide higher accuracy, while they are more time consuming and more vulnerable in convergence than lower-order schemes. On the other hand, high order schemes benefit more from the grid refinement. For instance, a size reduction factor of 2 can reduce the numerical error by a factor of 8 through a third-order scheme while a first-order one can only obtain a factor of 2. In general, the first-order discretization provides less accurate results with fewer convergence difficulties than the second-order one. In practice, it is suggested to apply a hybrid procedure, i.e. start the calculation with low order schemes (first-order upwind) providing initial conditions for high order schemes and end the final convergence study with at least second-order upwind scheme.

3.5 Mesh Generation

A mesh presents the physical model in a discrete form on which the governing equations can be resolved numerically. To obtain reliable CFD solutions, the mesh has to be sufficiently refined in the area where high gradients of fluid characteristics exist. This section discusses the usability of various mesh types. The applied hybrid meshes throughout the monograph, which have structured inflation layers around the profile and unstructured cells in the rest of the domain. Hybrid meshes are easier to generate and converge than structured meshes. Furthermore, hybrid meshes can better simulate the viscous effects than pure unstructured meshes.

3.5.1 Mesh Types

Mesh generation is one of the most important steps in CFD simulations, which significantly influences the calculation outcoming. However, concerning mesh properties, there are many factors that affect mesh quality and it is very time-costing to adjust

Table 3.2 ANSYS Fluent settings for mesh generation

Property	Setting
Solver	Pressure-based steady
Viscous model	$k - \omega$ SST
Pressure-velocity coupling	Coupled
Gradient	Least squares cell based
Pressure	Second order
Momentum	Second order upwind
Turbulent kinetic energy	Second order upwind
Specific dissipation rate	Second order upwind

those properties to get satisfactory results. In order to clarify how different mesh properties affect simulations of ship rudders, the classic validation profile NACA0012 is tested through commercial CFD code ANSYS Fluent 15.0. Throughout this work, the $k - \omega$ SST turbulence model is applied for steady incompressible viscous water with settings shown in Table 3.2.

According to the connectivity of elements, meshes are identified as structured, unstructured, and hybrid. As regular connectivity can be expressed as a two/three-dimensional array, the structured model is highly space-efficient. However, regular connectivity restricts the element type to quadrilateral in 2D and hexahedral in 3D, which inherently limits applications of the structured mesh for complex geometries. For simple geometries, a structured mesh may have better convergence and higher resolution than an unstructured mesh.

Instead of a single-block structured mesh, a block-structured or multiple-structured mesh is actually more widely applied. It is superior in computational efficiency to the unstructured mesh and more flexible in handling the complex geometries than the single-block structured mesh. The block-structured mesh has three common topologies, specifically C-mesh (Fig. 3.2), H-mesh (Fig. 3.3), and O-mesh (Fig. 3.4). Some

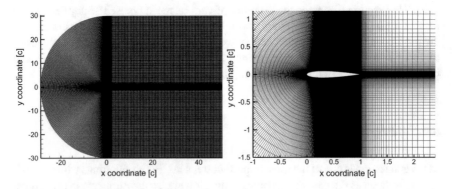

Fig. 3.2 An illustration of a block-structured C-mesh

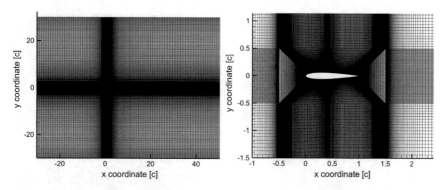

Fig. 3.3 An illustration of a block-structured H-mesh

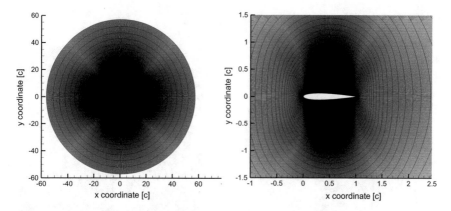

Fig. 3.4 An illustration of a block-structured O-mesh

modifications can be made based on these common topologies, which may lead to better accuracy, like O-H-Combination-Mesh (Fig. 3.5) and Modified-O-Mesh (Fig. 3.6). The choice of the mesh topology has a considerable influence on mesh quality [32], which depends on the domain geometry, the structure of the solution, and the topology in the adjacent block.

The irregular connectivity of unstructured meshes allows more freedom in element choices, typically triangles in 2D and tetrahedra in 3D. Compared to the structured mesh, the unstructured mesh is more suitable for complex configurations, but it can be highly inefficient. Aftosmis et al. [33] reported that unstructured triangular meshes are 50 times more expensive in both memory and time than structured quadrilateral meshes with nearly the same accuracy. Unstructured meshes also have advantages over structured meshes in mesh refinement and generation time.

In practice, hybrid meshes are applied instead of pure unstructured meshes. These hybrid meshes commonly consist of a structured portion near wall surfaces to capture the boundary layer and an unstructured portion that fills the domain. Thus, hybrid meshes inherit the advantages of both structured and unstructured meshes, such as

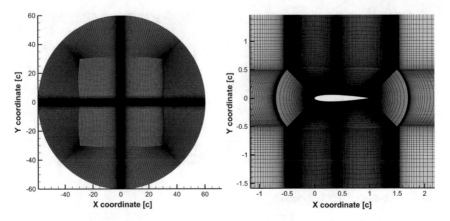

Fig. 3.5   An illustration of a block-structured O-H-1-mesh

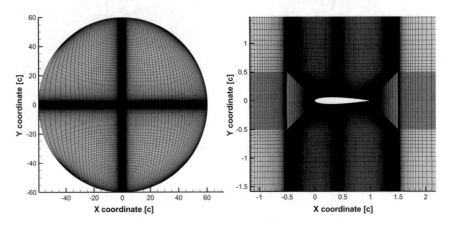

Fig. 3.6   An illustration of a block-structured O-H-2-mesh

good orthogonality to wall surfaces, suitability for mesh refinement, flexibility for complex geometries, and fast generation. Figure 3.7 demonstrates a hybrid triangle mesh, which has quadrilateral elements for the boundary layer and triangular elements for the remainder of the domain.

In literature, various choices of mesh types are used in the analysis of rudder-like objects as presented in Table 3.3. Van Nguyen and Ikeda [11, 34] applied hybrid triangle meshes for RANS simulations of fishtail rudders. Lutton [35] indicated that the O-mesh is superior to the C-mesh in determining the pressure coefficients in the vicinity of leading and trailing edges while the C-mesh shows a better resolution in the wake. Basha [36] compared the lift and drag coefficients through a C-mesh, an O-mesh, and a hybrid mesh with the same number of cells. The results obtained with the hybrid mesh are closest to the experiments. Comparing mesh dependent

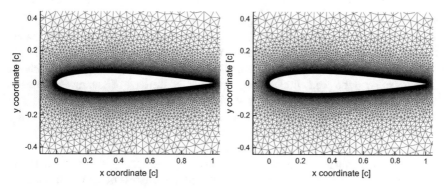

Fig. 3.7 An illustration of a hybrid triangle mesh

Table 3.3 Mesh types applied in literature

Literature	Mesh type
Stuck et al. [4]	Hybrid triangle mesh
Basha [36]	Block-structured C-mesh and O-mesh, hybrid triangle mesh
Wasberg and Reif [5]	Block-structured H-mesh
Narsipur et al. [37]	Hybrid triangle mesh
Tveiterås [38]	Block-structured O-mesh
Langley Research Center [39]	Block-structured C-mesh

deviation from measurements, Stuck et al. [4] confirmed the superior performance of a hybrid quad mesh to a hybrid triangle mesh.

To illustrate the impacts of the mesh type on solutions, simulations based on the C-mesh, the H-mesh, the O-mesh, and the hybrid triangle mesh are carried out. The domain shapes and sizes are illustrated in Figs. 3.2, 3.3, 3.4, and 3.7, respectively. A detailed discussion of the domain size is presented in Sect. 3.5.2. The inflation layers are kept the same, which have a chord-wise spacing of 1.78×10^{-3} c and a layer-wise growth rate of 1.1. These parameters are discussed in detail in Sect. 3.5.3.

Results based on various meshes are compared to the experimental benchmark [40] and above-mentioned CFD codes [39] in relative differences as presented in Table 3.4. All mesh types show a consistent underestimation of lift and an overestimation of drag. The underestimation of lift is mainly caused by numerical diffusion. All the CFD codes, i.e. CFL3D, FUN3D, NTS, and Fluent, are capable of predicting reasonable lift coefficients. But the predictions of drag coefficients are not as accurate as of that of lift coefficients, especially after stall.

The lift coefficient and drag coefficient are non-dimension forms of lift and drag respectively, while the former is mainly generated by the pressure difference between the two surfaces, the inviscid component, and the latter primarily contributed by the frictional shear stress, the viscous component. Since the lift can be determined to sufficient accuracy by simply integrating the pressure around the body, the viscous stresses need not be fully converged, while this is clearly not true of the drag [41]. What's more, the fact that lift is about an order of magnitude larger than drag makes it difficult to find a suitable level of convergence for both numerical results [41].

The grid generation strategies, numerical algorithms, and turbulence models all have impacts on the prediction of drag. Specifically, the lack of accurate turbulence modeling in regions of separated flow significantly influences the separated calculations (Holst, Computational Fluid Dynamics Drag Prediction-Results from the Viscous Transonic Airfoil Workshop). Therefore, numerical results of drag after stall are much more suspectable than that at smaller angles of attack.

In order to obtain more reliable and accurate drag coefficients, an advanced understanding of physics of flow around the rudder and the progress of applied turbulence models are needed, along with the development of discretization and numerical methods. Another reason for the overestimation of drag is the assumption of a fully turbulent flow along with the profile [8]. In fact, flow along with the profile always starts as laminar flow and transforms into turbulent flow, especially for low Reynolds numbers.

As Table 3.4 shows, the block-structured C-mesh has the best over-all performance in both lift and drag. The block-structured H-mesh is slightly worse than the C-mesh since the C-mesh has better orthogonality around the profile. The hybrid triangle mesh is comparable with C-mesh and H-mesh in lift prediction and much easier to generate. The O-mesh is very poor in the drag prediction as it fails to capture the wake. For all mesh types, the relative differences in drag increase as the angle of attack increases.

For aerodynamics, where both lift and drag are important to the performance, block-structured C-mesh is recommended. More advanced CFD methods, such as the Large Eddy Simulation (LES) and Direct Numerical Simulation (DNS), also help to improve the accuracy of predictions. On the other hand, rudder force in maneuvering is commonly estimated based on its normal force coefficient and its tangential force is neglected [42]. Considering the small contribution of drag to the rudder-induced side force, the accuracy of drag is less important than lift for maneuvering simulations. In that case, a hybrid triangle mesh can be a good choice, since it is easy to generate with sufficient accuracy of lift coefficients.

Considering the singularity that exists at the sharp trailing edge, we suggest applying a C-mesh for a single profile with a sharp trailing edge. For profiles with blunt edges or multiple profiles, an H-mesh could be alternative. Hybrid meshes can be applied for both cases, especially for complex geometries. The primary goal of this study is to determine suitable mesh properties to achieve accurate lift and drag coefficients. Therefore, discussions in the following sections are presented based on a block-structured C-mesh.

Table 3.4 A comparison of the results based on various mesh types to the experimental benchmark and other CFD codes in relative differences

Mesh type	Benchmark	ΔC_{D_0} (%)	$\Delta C_{L_{10}}$ (%)	$\Delta C_{D_{10}}$ (%)	$\Delta C_{L_{15}}$ (%)	$\Delta C_{D_{15}}$ (%)
Block-structured C-mesh	CFL3D	0.49	−1.45	4.85	−1.41	2.57
	FUN3D	0.62	−2.01	3.43	−1.68	0.04
	NTS	0.49	−1.33	3.60	−1.62	4.07
	Ladson	25.08	−2.37	24.43	−2.78	21.71
Block-structured H-mesh	CFL3D	1.56	−0.04	6.26	−3.92	4.66
	FUN3D	1.69	−0.61	4.82	−4.18	2.08
	NTS	1.56	−0.08	4.99	−4.13	6.19
	Ladson	26.41	−0.97	25.09	−5.26	24.19
Block-structured O-mesh	CFL3D	5.01	−5.88	98.76	−4.71	92.39
	FUN3D	5.14	−6.42	96.07	−4.96	87.66
	NTS	5.01	−5.77	96.38	−4.91	95.21
	Ladson	30.69	−6.76	133.97	−6.03	128.30
Block-structured O-H-1-mesh	CFL3D	1.52	−6.31	103.20	−4.24	56.80
	FUN3D	1.65	−6.85	100.45	−4.50	52.94
	NTS	1.52	−6.20	100.77	−4.45	59.09
	Ladson	26.36	−7.19	139.20	−5.57	86.06
Block-structured O-H-2-mesh	CFL3D	1.58	−4.69	58.90	−2.92	37.83
	FUN3D	1.70	−5.24	56.75	−3.19	34.44
	NTS	1.58	−4.58	56.99	−3.13	39.85
	Ladson	26.42	−5.58	87.05	−4.27	63.55
Hybrid triangle mesh	CFL3D	2.47	−1.45	13.83	−1.90	18.34
	FUN3D	2.60	−2.01	12.29	−2.17	15.43
	NTS	2.47	−1.33	12.47	−2.11	20.07
	Ladson	27.54	−2.38	34.00	−3.26	40.43

3.5.2 Domain Size

In order to obtain highly accurate solutions, the position of boundaries has to be discussed to demonstrate that the interior flow field is not disturbed. Thus, there is the need to study the domain size which not only minimizes the influences of the boundaries but also prevents an excessively large domain. Various domain sizes are applied in the literature in Table 3.3. Langley Research Center [39] argued that the far-field boundary locations affect the drag prediction more than the lift, proposing a far-field extent of $500\,c$.

The present study investigates the effects of the domain size on RANS solutions. Test parameters and related relative differences are given in Table 3.5. It is clear that the drag prediction is much more sensitive to the domain size than the lift prediction.

Table 3.5 A comparison of results with various domain sizes to the experimental benchmark in relative difference

Domain size (c)	No. cells	ΔC_{D_0} (%)	$\Delta C_{L_{10}}$ (%)	$\Delta C_{D_{10}}$ (%)	$\Delta C_{L_{15}}$ (%)	$\Delta C_{D_{15}}$ (%)
7.5/15	189596	25.38	−2.89	41.14	−2.90	38.56
15/30	272028	25.08	−2.50	28.57	−2.76	26.68
15/60	326735	25.38	−2.55	28.29	−2.85	26.31
30/30	378480	25.08	−2.49	26.48	−2.80	24.87
30/60	456075	25.08	−2.37	23.43	−2.78	21.71
60/120	985395	25.23	−2.28	21.14	−2.79	19.52

With an increase of the domain size, the accuracy of the drag coefficient is more significantly improved than the lift. However, further expansion of the domain size may dramatically increase the number of cells and computation time.

A domain size of 30 c around and 60 c after the profile achieves a good balance in both accuracy and efficiency. This domain size is sufficiently large to obtain accurate estimates of lift and drag. The number of cells is still manageable by common desktop computers. Thus, it is applied for the remainder of the experiments in this paper, since the objective of this study has an interest in both lift and drag. Table 3.5 indicates that more accurate predictions can be obtained with larger domains. Thus, we suggest applying a large domain whenever computation power and time are available. If the lift is the only purpose, a small domain size of 15 c around and 30 c after the profile may be more favorable as it requires less computation power and time than a large domain.

3.5.3   Grid Independence

CFD results must be independent of the mesh resolution. Commonly a fine mesh helps to improve the calculation accuracy. Not checking mesh independence is one of the common error sources in CFD simulations. The same meshing strategy is only applicable to similar problems. At least one convergence study has to be done for each new problem [2]. However, a mesh may also be unnecessarily refined resulting in a waste of cells and computation time. Wasberg and Reif [5] showed that the accuracy of the drag coefficients depends on the domain size and the turbulent flow conditions. Furthermore, the domain size mainly affects the prediction of the drag while its impacts on the lift are less significant. Comparing to wind-tunnel tests [40].

After determining the mesh type, the domain size, and the test Reynolds number, a raw mesh should be ready for the mesh independence study. The common procedure is to carry out simulations on an initial mesh with a residual error in a range of 1×10^{-4} to 1×10^{-5}. After that, refine the mesh globally to around 2 or $\sqrt{2}$ times the initial mesh. Next, run simulations with the refined mesh and compare solutions obtained from the coarse mesh and the fine mesh. Repeat the refinement until the results do not significantly change with a finer mesh. The mesh independence study requires at least 3 solutions to evaluate the convergence of certain inputs [43]. Considering the computation time, it is always better to use the smallest number of cells.

Instead of rough global refinement of cells [44], two main impact factors on mesh fineness and the required number of cells, i.e. the chord-wise spacing on the profile, which determines the aspect ratio of the boundary layer elements, and the layer-wise growth rate, which describes the nodes distribution along radial directions of the boundary layer meshes, are illustrated with simulations in this monograph. This parametric mesh independence study is chosen, because optimization on individual parameters addresses the crucial impact factor of mesh quality more flexibly, avoiding a waste of cells [4]. In fact, global refinement is more frequently applied for unstructured meshes than structured meshes.

3.5.3.1 Chordwise Spacing

The chord-wise spacing is the distance between two nodes along with the profile. The first mesh height is commonly fixed to obtain sufficient resolution in the boundary layers. Small chord-wise spacing can decrease the aspect ratio of the cells and increase the orthogonal quality. To focus on the impacts of the chord-wise spacing on solutions, the test meshes are spread with a specific expansion ratio of 1.1. Case settings are presented in Table 3.6 including the maximum chord-wise spacing and the total number of cells.

Table 3.6 demonstrates that the reduction of chord-wise spacing does not consistently improve the accuracy of the solutions. As the chord-wise spacing decreases, the number of elements with large aspect ratios in the wake increases, which introduces difficulties of convergence. A chord-wise spacing of 1.78×10^{-3} reaches the limitation of improvements through the reduction of the spacing, after which the number of cells dramatically increases while the accuracy drops. As the resolution

Table 3.6 A comparison of results with various chord-wise spacing to the experimental benchmark in relative differences

Spacing (c)	No. cells	ΔC_{D_0} (%)	$\Delta C_{L_{10}}$ (%)	$\Delta C_{D_{10}}$ (%)	$\Delta C_{L_{15}}$ (%)	$\Delta C_{D_{15}}$ (%)
4.46×10^{-3}	151076	25.08	−2.97	32.67	−4.77	39.14
3.57×10^{-3}	167186	25.08	−2.79	31.81	−3.71	32.30
2.23×10^{-3}	201912	25.23	−2.60	29.43	−3.23	29.47
1.78×10^{-3}	272028	25.08	−2.50	28.57	−2.76	26.68
8.92×10^{-4}	484266	25.23	−2.62	29.52	−3.41	31.60
4.46×10^{-4}	889272	–	–	–	–	–

of mesh increases, convergence problems also occur due to the fact that small flow features are being solved, for instance, the last case of 4.46×10^{-4} c diffuses after hundreds of iterations.

3.5.3.2 Layer-Wise Growth Rate

The layer-wise growth rate is the last mesh property to investigate in this paper. It is the ratio of distances between adjacent mesh points along a mesh line. The mesh growth rate here is specified along the direction from the profile to the far-field. As discussed in the last section, a chord-wise spacing of 1.78×10^{-3} c is chosen to explore the effects of the layer-wise growth rate. The case settings are given in Table 3.7.

Table 3.7 indicates that the layer-wise growth rate affects the drag more than the lift. The decrease of the growth rate improves the drag prediction to a value of 1.1, after which the accuracy goes down while the number of cells increases significantly leading to a high computation cost. A growth rate of 1.03 encounters convergence problems after an angle of attack of 5°. The present results show that a growth rate of 1.1 is well balanced in accuracy and computation time.

Comparing the results in Tables 3.6 and 3.7, the findings provide evidence that simulations are more sensitive to the chord-wise spacing than layer-wise growth rate. Finally, a mesh with a chord-wise spacing of 1.78×10^{-3} c and a layer-wise growth rate of 1.1 leads to mesh independent results. We also proposed to apply these parameters as guidance for unit length profile simulations.

Figure 3.8 illustrates the 2D meshes for a NACA 0012 profile which are used for the validation of the 2D RANS method in Sect. 4.1. Meshes for other tested profiles are generated with the same strategy. The test Reynolds number is 6×10^6. Above this value, small impacts of Re on the lift and drag coefficients are found (Sect. 2.2.1). Corresponding to the Re, the inflow speed is 6.0289 m/s for a rudder model with chord length of 1 m. The inflow is kept normal to the inlet boundary. Furthermore, the angle of attack is configured by rotating the profile rather than modifying the direction of the inflow.

Table 3.7 A comparison of results with various layer-wise growth rates to the experimental benchmark in relative differences

Growth rate	No. cells	ΔC_{D_0} (%)	$\Delta C_{L_{10}}$ (%)	$\Delta C_{D_{10}}$ (%)	$\Delta C_{L_{15}}$ (%)	$\Delta C_{D_{15}}$ (%)
1.2	190920	24.46	−2.58	30.00	−2.74	28.66
1.15	216000	24.92	−2.49	29.24	−2.63	27.27
1.1	272028	25.08	−2.50	28.57	−2.76	26.68
1.05	438400	26.00	−2.70	28.29	−3.26	27.43
1.03	636800	26.15	−2.82	40.29	–	–

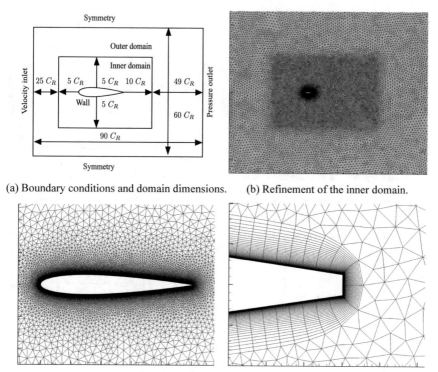

(a) Boundary conditions and domain dimensions.

(b) Refinement of the inner domain.

(c) Structured meshes around the profile.

(d) Connection of the meshes.

Fig. 3.8 2D mesh topology, boundary conditions, and domain dimensions

2D meshes are implemented in a rectangular domain of 60 C_R in width and 90 C_R in length. The profile is located towards the front of the domain, with 30 C_R to left, top and bottom, and 59 C_R to right respectively. This domain is sufficiently large to minimize the influence of the boundary locations on the rudder hydrodynamics while not excessively large. As illustrated in Fig. 3.1, an inner domain is defined to improve the accuracy of the solution and capture the wake area around the rudder.

3.5.4 3D Domain and Meshes

Similar to the 2D meshes, the presented 3D meshes are set up in a cuboid domain. The dimensions of the domain, the topology of the mesh, and the applied boundary conditions are illustrated in Fig. 3.9, where d_R is the clearance between the rudder tip and the bottom of the waterway. Additionally, d_R is non-sensationalized as $d'_R = d_R/B_R$. An inner domain is built to refine the necessary cells around the rudder. The velocity inlet condition defines the inflow velocity according to the tested Reynolds

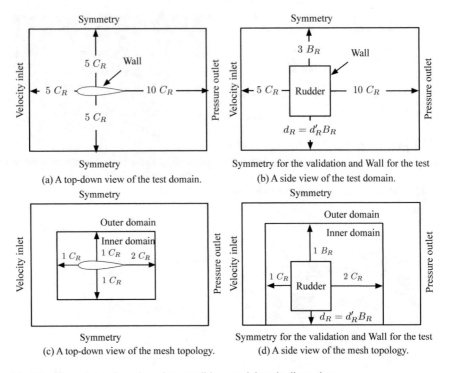

Fig. 3.9   3D mesh topology, boundary conditions, and domain dimensions

number (6×10^6 in water) or the value specified in the validation experiment (20 m/s in the air). The pressure outlet condition sets the pressure as constant and atmospheric.

3.5.5   Grid Independence

A grid independence test shows that the solutions are independent of the change in grids. This monograph refines meshes based on the chord-wise element size along with the rudder profile, which determines the aspect ratio of the boundary layer elements. This parametric refinement addresses the crucial impact factors on the mesh quality and avoids the waste of cells by overall refinement [44]. Wasberg and Reif [5] showed that the accuracy of the drag coefficients depends on the domain size and the turbulent flow conditions. Furthermore, the domain size mainly affects the prediction of the drag while its impacts on the lift are less significant. Comparing to wind-tunnel tests [40]. An example of the grid independence is performed with a NACA 0012 profile at the angle of attack of 10° as shown in Fig. 3.10.

As the first cell height is 4.46×10^{-6} C_R, the chord-wise element size is tested in a range of 2.68×10^{-3} to 2.23×10^{-4} C_R. The corresponding range of aspect ratios

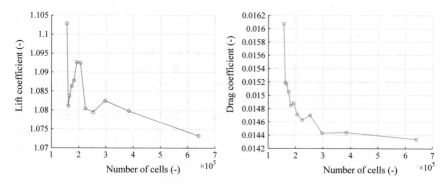

Fig. 3.10 Grid independence study of the lift and drag coefficients at an angle of attack of 10°

is 600 to 50 at an interval of 50. Figure 3.10 confirms that the accuracy of the drag coefficient is more sensitive to the number of cells than that of the lift coefficients. Through the mesh independence study, a mesh of 400 000 cells is deemed to be sufficient to keep the results independent from the grid for a 2D single rudder. Similar procedures are taken for twin-rudder and quadruple-rudder configurations in 2D and single-rudder configurations in 3D. More specifically, the aspect ratio of the first layer of the structured mesh around the profile in 2D and 3D is about 100 and the chord-wise size is about $4.46 \times 10^{-5} \, C_R$. For these three cases, the number of cells at which mesh independence is achieved are 700 000, 1 200 000, and 3 000 000 respectively.

3.5.6 Concluding Remarks of Mesh Generation

A good mesh is the prerequisite of solid CFD simulations. A mesh generation starts with the determination of the mesh type, continues with the property settings, and finalizes with the grid independence procedure. A good mesh should be generated in such a way that little human effort is requested. Thus, the results can be compared to other results without relying on the experts' skills. The control properties of a mesh should be carefully calibrated to prevent extensive domain size, grid refinement, and computational time. Furthermore, the mesh must be independent of the iteration steps.

3.6 Summary

Take the Reynolds-Averaged Navier–Stokes method as an example, the applications of Computational Fluid Dynamics methods to simulations on rudder hydrodynamics have been discussed in this chapter. Following governing equations in RANS studies in Sect. 3.1, the selections of key properties during the procedure are presented based on characteristics of rudder hydrodynamics. A trade-off between computation accuracy and efficiency needs to be balanced while conducting rudder simulations, which determines specific simulation strategies in practice.

The impacts of turbulence models and numerical solvers are discussed in Sects. 3.2, 3.3, and 3.4. Clear understandings of physical conditions of rudders are required for appropriate choices. As for mesh generations, systematic investigations on related properties, like mesh types, domain sizes, and nodes distributions are recommended for typical objects if necessary as discussed in Sect. 3.5. Since it is difficult to quantify the impacts of properties above in strict theoretical forms considering various rudder types and working conditions, a step-by-step investigation can provide relatively reliable results by parameter-to-parameter comparisons.

References

1. Molland AF, Turnock SR (2007) Marine rudders and control surfaces: principles, data, design and applications, 1st edn. Elsevier Butterworth-Heinemann, Oxford
2. MARNET-CFD (2002) Best practice guidelines for marine applications of computational fluid dynamics. Retrieved from https://pronet.atkinsglobal.com/marnet/guidelines/guide.html. Accessed 3 Dec 2014
3. Versteeg HK, Malalasekera W (2007) An introduction to computational fluid dynamics: the finite volume method, 2nd edn. Pearson Education, London
4. Stuck A, Turnock S, Bressloff N (2004) An evaluation of the RANS method for the prediction of steady ship rudder performance compared to wind tunnel measurements. Technical report, University of Southampton, Southampton, UK
5. Wasberg CE, Reif BAP (2010) Hydrodynamical simulations in fluent. Technical report, Norwegian Defence Research Establishment (FFI), Kjeller, Norway
6. Krasilnikov V, Ponkratov D, Crepier P (2011) A numerical study on the characteristics of the system propeller and rudder at low speed operation. In: 2nd international symposium on marine propulsors, Hamburg, Germany
7. Castro AM, Carrica PM, Stern F (2011) Full scale self-propulsion computations using discretized propeller for the KRISO container ship KCS. Comput Fluids 51:35–47
8. Eleni DC, Athanasios TI, Dionissios MP (2012) Evaluation of the turbulence models for the simulation of the flow over a national advisory committee for aeronautics (NACA) 0012 airfoil. J Mech Eng Res 4(3):100–111
9. Karim MM, Ahmmed MS (2012) Numerical study of periodic cavitating flow around NACA 0012 hydrofoil. Ocean Eng 55:81–87
10. Broglia R, Dubbioso G, Durante D, Di Mascio A (2013) Simulation of turning circle by CFD: analysis of different propeller models and their effect on manoeuvring prediction. Appl Ocean Res 39:1–10
11. Van Nguyen T, Ikeda Y (2013) Hydrodynamic characteristic of rudder sections with high lift force. J Jpn Soc Nav Arch Ocean Eng 403–406
12. Van Nguyen T, Ikeda Y (2014) Hydrodynamic characteristic of rudder sections with high lift force, part 2: the wedge tail shapes. J Jpn Soc Nav Arch Ocean Eng 171–174
13. Van Nguyen T, Ikeda Y (2014) Hydrodynamic characteristic of rudder sections with high lift force, part 3: the trailing edge with flat plate. J Jpn Soc Nav Arch Ocean Eng 171–174
14. Shenoi RR, Krishnankutty P, Selvam RP (2014) Study of manoeuvrability of container ship by static and dynamic simulations using a RANSE-based solver. Ships Offshore Struct 1–19

15. Wang HZ, Zou ZJ (2014) Numerical study on hydrodynamic interaction between a berthed ship and a ship passing through a lock. Ocean Eng 88:409–425
16. Van Nguyen T, Ikeda Y (2015) Development of marine high lift performance rudders with wedge tails: effect of Reynolds number on hydrodynamic forces. In: International conference on transportation mechanical engineering and sustainable development, Danang, Vietnam
17. Tezdogan T, Demirel YK, Kellett P, Khorasanchi M, Incecik A, Turan O (2015) Full-scale unsteady RANS CFD simulations of ship behaviour and performance in head seas due to slow steaming. Ocean Eng 97:186–206
18. Badoe CE, Phillips AB, Turnock SR (2015) Influence of drift angle on the computation of hull-propeller-rudder interaction. Ocean Eng 103:64–77
19. Broglia R, Dubbioso G, Durante D, Di Mascio A (2015) Turning ability analysis of a fully appended twin screw vessel by CFD, part I: single rudder configuration. Ocean Eng 105:275–286
20. Dubbioso G, Durante D, Mascio AD, Broglia R (2016) Turning ability analysis of a fully appended twin screw vessel by CFD, part II: single vs. twin rudder configuration. Ocean Eng 117:259–271
21. Bhattacharyya A, Krasilnikov V, Steen S (2016) Scale effects on open water characteristics of a controllable pitch propeller working within different duct designs. Ocean Eng 112:226–242
22. Spalart PR, Allmaras SR (1994) A one equation turbulence model for aerodynamic flows. Rech Aerosp 5–21
23. Launder BE, Spalding DB (1974) The numerical computation of turbulent flows. Comput Methods Appl Mech Eng 3(2):269–289
24. Quérard A, Temarel P, Turnock SR (2008) Influence of viscous effects on the hydrodynamics of ship-like sections undergoing symmetric and anti-symmetric motions using RANS. In: 27th international conference on offshore mechanics and arctic engineering (OMAE 2008), Estoril, Portugal, vol 5, pp 683–692
25. Date JC (2001) Performance prediction of high lift rudders operating under steady and periodic flow conditions. PhD thesis, University of Southampton
26. Yakhot V, Orszag SA, Thangam S, Gatski TB, Speziale CG (1992) Development of turbulence models for shear flows by a double expansion technique. Phys Fluids A: Fluid Dyn 4(7):1510–1520
27. Shih TH, Liou WW, Shabbir A, Yang Z, Zhu J (1995) A new k-ε eddy viscosity model for high Reynolds number turbulent flows. Comput Fluids 24(3):227–238
28. Wilcox DC (1988) Reassessment of the scale-determining equation for advanced turbulence models. AIAA J 26(11):1299–1310
29. Wilcox DC (2008) Formulation of the kw turbulence model revisited. AIAA J 46(11):2823–2838
30. Menter FR (1994) Two-equation eddy-viscosity turbulence models for engineering applications. AIAA J 32(8):1598–1605
31. Kelecy FJ (2008) Coupling momentum and continuity increases CFD robustness. ANSYS Advant 2(2):49–51
32. Liseikin VD (2009) Grid generation methods, 2nd edn. Springer, Berlin
33. Aftosmis M, Gaitonde D, Tavares TS (1995) Behavior of linear reconstruction techniques on unstructured meshes. AIAA J 33(11):1–51
34. Van Nguyen T, Ikeda Y (2014) Development of fishtail rudder sections with higher maximum lift coefficients. In: 24th international ocean and polar engineering conference, Busan, Korea, pp 940–947
35. Lutton MJ (1989) Comparison of C-and O-grid generation methods using a NACA 0012 airfoil. Master thesis, Air Force Institute of Technology
36. Basha W (2006) Accurate drag prediction for transitional external flow over airfoils. Master thesis, Concordia University
37. Narsipur S, Pomeroy BW, Selig MS (2012) CFD analysis of multielement airfoils for wind turbines. In: 30th AIAA applied aerodynamics conference, Louisiana, New Orleans, pp 1–18

38. Tveiterås V (2011) Numerical study of the interaction of flow over two airfoils in relative motion. Master's thesis, Norwegian University of Science and Technology
39. Langley Research Center (2014) Turbulence modeling resource. Retrieved from http://turbmodels.larc.nasa.gov/naca0012_val.html. Accessed 10 Sept 2014
40. Ladson CL (1988) Effects of independent variation of mach and reynolds numbers on the low-speed aerodynamic characteristics of the NACA 0012 airfoil section. Technical report, Langley Research Center, Hampton, Virginia, USA
41. Roberts TW (1990) Prospects for drag prediction using computational fluid dynamics. In: Aerospace technology conference and exposition, Long Beach, CA, USA
42. Yasukawa H, Yoshimura Y (2015) Introduction of MMG standard method for ship maneuvering predictions. J Mar Sci Technol 20(1):37–52. https://doi.org/10.1007/s00773-014-0293-y
43. International Towing Tank Conference (2008) ITTC recommended procedures and guidelines 7.5-03-01: uncertainty analysis in CFD verification and validation
44. Stern F, Wilson RV, Coleman HW, Paterson EG (1999) Verification and validation of CFD simulations. Technical report, Iowa Institute of Hydraulic Research, Iowa City, USA

Chapter 4
Hydrodynamic Characteristics of Ship Rudders

As discussed in Chap. 2, inland vessels have more complex rudder configurations than seagoing ships. Currently, the rudder profile, the number of rudders, and the spacing between rudders are not considered in the empirical methods for rudder force calculation. To analyze the impacts of the rudder configurations on inland vessel maneuverability, the hydrodynamic characteristics, for instance, lift and drag coefficients, of each configuration need to be calculated specifically. Considering the scaling effects and high cost of model tests, one of the Computational Fluid Dynamic (CFD) methods, the Reynolds-Averaged Navier-Stokes (RANS) method is applied to analyze the hydrodynamics of rudders in different configurations.

Section 4.1 validates the presented RANS methods against experimental and numerical test data. Section 4.2 uses the 2D RANS method to study the impacts of the Reynolds numbers, the rudder profiles, the spacing between twin rudders, and the spacing among quadruple rudders. Section 4.3 analyzes the impacts of the effective aspect ratio and shallow water on rudder hydrodynamics through 3D RANS simulations. Based on the obtained hydrodynamic characteristics of different rudder configurations, regression formulas are propose in Sect. 4.4 and applied to maneuvering simulations in Chaps. 6 and 8. Finally, Sect. 4.5 concludes this chapter.

4.1 Validation of RANS Methods

4.1.1 Validation of the 2D RANS Method

To validate the RANS method, a classic validation profile NACA 0012 is analyzed under angles of attack in a range of 0–5° at an interval of 1°. The lift and drag coefficients are compared to one experimental data set [1] and three independent

© Springer Nature Switzerland AG 2020
J. Liu, *Mathematical Modeling of Inland Vessel Maneuverability Considering Rudder Hydrodynamics*, https://doi.org/10.1007/978-3-030-47475-1_4

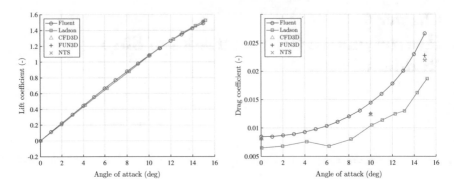

Fig. 4.1 Comparison of the 2D Fluent results to the numerical and experimental tests

CFD results [2], which are CFL3D (NASA LaRC, USA), FUN3D (NASA LaRC, USA), and NTS (NTS, Russia), as shown in Fig. 4.1. These three benchmark CFD cases were carried out with structured meshes in a domain of 500 C_R around the profile. Each CFD case has three data points at angles of attack of 5°, 10°, and 15°.

The comparison in Fig. 4.1 shows that the 2D RANS method with hybrid meshes predicts the lift coefficient well but overestimates the drag coefficient. The difference in the lift prediction is mainly due to numerical diffusion. The applied 2D RANS method has a larger overestimation of the drag coefficients than the CFD benchmarks because hybrid meshes introduce larger discretization errors than the structured meshes applied in the benchmark cases. The accuracy of the drag coefficient can be improved by using fully-structured meshes, a larger domain, or more advanced CFD methods like Large Eddy Simulations or Direct Numerical Simulations (DNS). However, these improvements are quite expensive in computation resources and time. Since the drag coefficient is a relatively small value, which contributes to the normal force coefficient much less than the lift coefficient, the inaccuracy of the drag coefficients is deemed for angles in the range of 0–15°.

Due to the availability of the experimental data, the validation was only performed for 0–15°. For angles of attack larger than 15°, uncertainty is caused by less validation and the strong flow separation, especially for the wedge-tail series. However, these large angles hardly happen at service speed operations, so their impacts on fuel consumption are small. To minimize these uncertainties, the regression formulas of the rudder hydrodynamics to be developed in Sect. 4.4 are based on the data points in the range of 0–10°. The results are shown in the full range of the applied rudder angles (0–35°) for maneuvering simulations to show the tendency of the coefficients, but the uncertainty of the results at the large angles of attack should be noticed.

The presented validation proves the usability of the RANS method for the NACA series. For the IFS series, only low-Reynolds-number results were found in the literature, which was given by [3]. The validation for the IFS series was not performed with these data which may be affected by the low Reynolds number. Since both the NACA and IFS series are well-streamlined profiles, it is reasonable to assume the method is applicable for the IFS series. Yet, no accurate geometry nor validation data for the wedge-tail rudders are available in the literature. The tail shape may cause stronger flow separation than the NACA and IFS series at large angles of attack while not significantly change the stall angle, which causes uncertainty of using the presented 2D RANS method.

4.1.2 Validation of the 3D RANS Method

To validate the 3D RANS method, a model is set up according to the wind-tunnel tests of a spade NACA 0020 rudder carried out by [4]. The chord length and the span of the test rudder are 0.667 m and 1 m respectively. The geometric aspect ratio is 1.5. As the gap between the rudder tip and the bottom of the wind tunnel is very small (2.5 mm), the effective aspect ratio of the rudder is 3.0. The wind tunnel is 3.5 m long, 2.5 m wide, and 2.5 m high. The inflow velocity is 20 m s^{-1}. The working fluid is air. The 3D RANS method to validate is configured according to the wind-tunnel tests, except that the domain is larger than the wind tunnel as shown in Fig. 3.9. The larger domain is chosen to minimize the influence of the location of the non-physical boundaries on the RANS results.

The rudder tip is connected with the bottom of the domain for simplicity. The viscous effect of the bottom is not accounted for in the validation case, but it is accounted for the simulations for the shallow water effect (Sect. 4.3). Furthermore, the inflow is air at 20 m s^{-1} in the validation case while it is water at a Reynolds number of 6×10^6 for other 3D simulations. Figure 4.2 compares the results of the 3D RANS method and the wind-tunnel tests. In general, the absolute relative differences of the lift, drag, and normal force coefficients are 8 %–11 %, 6 %–28 %, and 4 %–12 % respectively. The accuracy of the method can be improved by changing the turbulence model, increasing the number of cells, and enlarging the domain. However, the computational time may increase significantly.

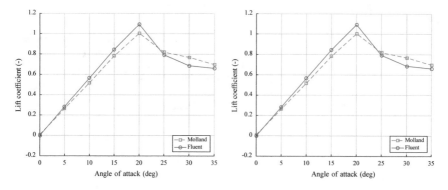

Fig. 4.2 Comparison of the 3D Fluent results to the experimental tests

4.1.3 Concluding Remarks of Validation of the RANS Methods

The applied CFD methods are validated with results from other CFD tests and public available experimental tests based on NACA profiles. Afterward, the validated method is applied to other profiles that do not have experimental results open to the public. The method is conducted according to best practices. Furthermore, other profiles are not dramatically different from the NACA profile. Thus, it is reasonable to use the validate method to research cases without validation data, which is not available.

4.2 2D RANS Study on Rudder Hydrodynamics

With the validated 2D RANS method, this section studies the hydrodynamic characteristics of single-rudder, twin-rudder, and quadruple-rudder configurations. For each rudder, the force conventions are illustrated in Fig. 4.3. In this monograph, counter-clockwise angles are taken as positive.

The relative positions of multiple rudders are defined by the lateral spacing between the rudder stocks, which are shown in Fig. 4.4. For twin-rudder configurations, the spacing between the two rudder stocks (y_{TR}) is normally around the value of the propeller diameter (D_P). Additionally, for inland vessels, the rudder chord length (C_R) is commonly not much different from D_P. Therefore, C_R is used as the non-dimensional factor for spacing. A quadruple-rudder configuration can be regarded as a combination of two twin-rudder units with specified spacing between the two inner stocks (y_{TU}). When y_{TR} and y_{TU} are large enough, no interaction effect is expected.

Fig. 4.3 Rudder force conventions in RANS simulations

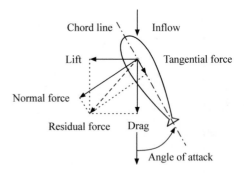

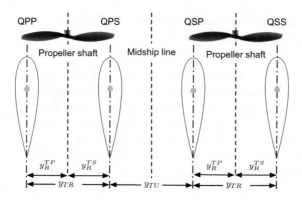

Fig. 4.4 Conventions of twin-rudder and quadruple-rudder configurations

Table 4.1 Test configurations for studies on rudder hydrodynamics in 2D

Section	Profile	Re $(-)$	y_{TR} (C_R)	y_{TU} (C_R)
Section 4.2.1	NACA 0012	2×10^5 to 1×10^7	–	–
Section 4.2.2	NACA 0015, NACA 0020, NACA 0025	6×10^6	0.5, 1.0	–
	IFS58 TR15, IFS61 TR25, IFS62 TR25			
	Wedge-tail 0015, Wedge-tail 0020, Wedge-tail 0025			
Section 4.2.3	NACA 0018	6×10^6	0.4, 0.5, 0.6, 0.7 0.8, 0.9, 1.0	–
Section 4.2.4	NACA 0015, NACA 0020, NACA 0025	6×10^6	0.5, 0.6	–
	IFS58 TR15, IFS61 TR25, IFS62 TR25		0.7, 0.8	
	Wedge-tail 0015, Wedge-tail 0020, Wedge-tail 0025		0.9, 1.0	
Section 4.2.5	NACA 0018	6×10^6	0.5, 0.75, 1.0	0.5, 1.0, 1.5

The presented test configurations are listed in Table 4.1. The impacts of the Reynolds number on rudder hydrodynamics (Sect. 4.2.1), the impacts of profiles on the performance of single rudders (Sect. 4.2.2), the impacts of the spacing on twin rudders (Sect. 4.2.3), the impacts of the profile on twin rudders (Sect. 4.2.4), and the impacts of the spacing on quadruple rudders (Sect. 4.2.5) are studied respectively.

4.2.1 Impacts of Reynolds Numbers on Rudder Hydrodynamics

The Reynolds number (Re) is defined as the ratio of inertial forces to viscous forces. It represents the similarity between flow patterns and determines the optimal chordwise and layer-wise mesh sizes. Due to the limited model size and capacity of the test facility, Re of tests in wind tunnels and towing tanks has to be scaled. For CFD methods, full-scale simulations with realistic Re are possible but expensive due to the need for a fine mesh and a large domain. Experimental results of aerodynamics with a small Mach number can be taken as validation results of incompressible water simulations, as the compressibility effects of a fluid with a Mach number smaller than 0.2 are small.

Ladson [1] observed impacts in the lift-curve slope, the maximum lift coefficient, and the maximum lift-drag ratio with increasing Re. Common benchmark wind tunnel tests are carried out at Re in the range of 1×10^5 [3] to 1×10^7 [1]. Nowadays, low-Reynolds-number RANS analysis is still challenging [5, 6] and high-Reynolds-number simulations may be expensive in computation time. The present work tests the NACA0012 profile at Re in a range of 2×10^5 to 1×10^7 as shown in Fig. 4.5.

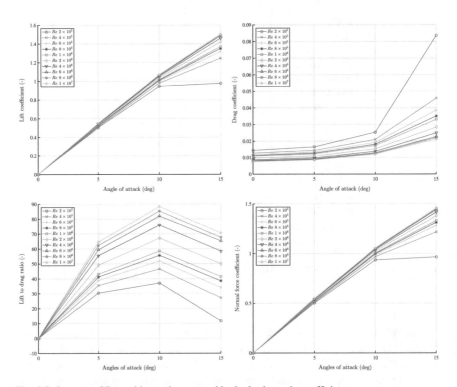

Fig. 4.5 Impacts of Reynolds numbers on rudder hydrodynamic coefficients

These tests are performed with the same grid, boundary conditions, and turbulence model as introduced in Chap. 3.

The lift curve rises with impacts in Re while the drag curve decreases. Compared to the lift coefficients, the drag coefficients are more sensitive to changes in Re. The drag coefficient under $15°$ at Re of 1×10^7 is about a quarter of the value at Re of 2×10^5, whereas the lift coefficient is 1.5 times larger. The differences of lift and drag between low and high Re increase with an increasing angle of attack. Consistent with findings by [1, 7], a Reynolds number of 6×10^6 can be considered as a threshold value above which little variation may be found.

4.2.2 Impacts of Profiles on Single-Rudder Hydrodynamics

2D open water hydrodynamic coefficients of various rudder profiles, which are the same as the hydrodynamic characteristics of 3D rudders with infinite aspect ratios in open water, are obtained using the 2D RANS method that is validated in Sect. 4.1.1. Figure 4.6 illustrates the impacts of the profile on rudder efficiency (lift to drag ratios) and effectiveness (normal force coefficients). These RANS results show that the NACA series is most efficient while the wedge-tail series is most effective. However, the NACA series and the wedge-tail series are least effective and efficient respectively. It not possible to achieve high efficiency (high lift to drag ratios) and high effectiveness (high lift coefficient at a given angle of attack) at the same time. The IFS series, which is initially designed for ships [3], achieves a balance of efficiency and effectiveness.

Figure 4.6 also shows that thinner NACA profiles have higher lift and drag coefficients than thicker NACA profiles, resulting in higher normal force coefficients that are dominated more by the lift coefficients than the drag coefficients. The thickness of the NACA profile does not significantly influence the stall angle. Unlike the NACA series, the thinner wedge-tail and IFS profiles have lower lift coefficients than the thicker ones. Besides, thinner wedge-tail and IFS profiles have higher drag coefficients. Due to the change of the profile thickness, the IFS series has a significant change in the lift coefficient and the stall angle. Moreover, a change of the tail thickness extends the stall angles and raises the lift to drag ratios of the IFS profiles. Compared to the NACA and wedge-tail series, thicker IFS profiles have smaller stall angles.

Summarizing, various rudder profiles have different hydrodynamic characteristics, which further affect ship maneuvering performance. The NACA series is most economical. Thus, it is widely applied to ships without critical maneuvering requirements. The wedge-tail series is most effective but least efficient. Wedge-tail rudders are suggested for ships that need exceptional maneuvering performance, sail in constrained waterways, or have limited rudder area, for instance, inland vessels. The IFS series can be a good choice when trying to balance efficiency and effectiveness.

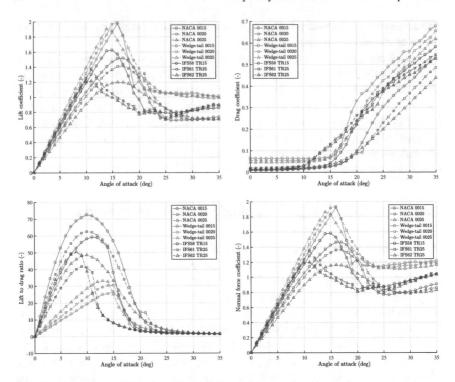

Fig. 4.6 2D open water hydrodynamic coefficients of the tested rudder profiles

4.2.3 Impacts of Spacing on Twin-Rudder Hydrodynamics

The spacing between twin rudders (y_{TR}) affects the pressure distribution around the twin rudders and thus changes the hydrodynamic characteristics. To analyze the impacts of y_{TR} on twin-rudder hydrodynamics, single NACA 0018 rudder and twin NACA 0018 rudders with y_{TR} in the range of 0.4–1.0 C_R at an interval of 0.1 C_R are tested. The test range of angles of attack is 0–35°. Figure 4.7 illustrates the pressure distributions of twin rudders with different y_{TR} at an angle attack of 10°.

A clear low-pressure region lies between the two rudders. As y_R increases, the pressure difference between the two sides of the port side rudder increases while that of the starboard side rudder decreases, which leads to an increase in the port side lift coefficient and a decrease in the starboard side lift coefficient. The pressure at the leading edge of the port side rudder is decreased while that of the starboard side rudder is increased. This phenomenon explains the changes in the drag coefficients of each rudder. Figure 4.8 compares the lift and drag coefficients of the single rudder and each rudder in the twin-rudder configurations with various y_{TR}.

Compared to the hydrodynamic coefficients of the single rudder, the stall angles of both the port side and the starboard side rudders are increased. Especially, for the starboard side rudders, the sharp decrease of the lift coefficient after the stall angle

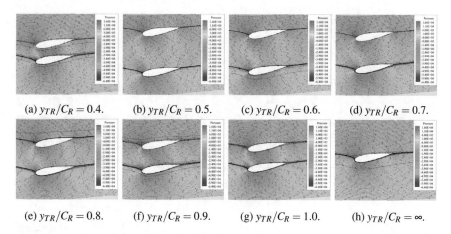

(a) $y_{TR}/C_R = 0.4$.   (b) $y_{TR}/C_R = 0.5$.   (c) $y_{TR}/C_R = 0.6$.   (d) $y_{TR}/C_R = 0.7$.

(e) $y_{TR}/C_R = 0.8$.   (f) $y_{TR}/C_R = 0.9$.   (g) $y_{TR}/C_R = 1.0$.   (h) $y_{TR}/C_R = \infty$.

Fig. 4.7 Pressure distributions with stream traces of twin NACA 0018 rudders at an angle of attack of $10°$ with various y_{TR}

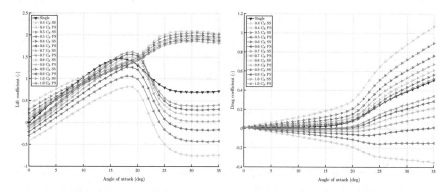

Fig. 4.8 Lift and drag coefficients of a single NACA 0018 rudder and either side of twin NACA 0018 rudders with various y_{TR}

does not appear. The lift slopes of the port side and starboard side rudders are roughly the same but smaller than that of the single-rudder case. In general, the starboard side rudder has a higher lift coefficient than the port side rudder. However, with y_{TR} of $0.9\,C$ and $1.0\,C$, the port side rudder under angles of attack in the range of $10°$ to $16°$ has smaller lift coefficient than those of the starboard side rudder.

The drag coefficients of the starboard side and the port side rudders in twin-rudder configurations are approximately symmetric with the drag coefficient of the single rudder. The drag coefficients at large angles of attack (larger than $15°$) are more sensitive to the change of y_{TR} because the frictional drag is dominant at small angles of attack while the pressure drag is dominant at large angles of attack. It should be noted that the port side rudder may have a negative drag coefficient, which means generating thrust instead of resistance while the starboard side rudder has much higher drag coefficient than the single rudder.

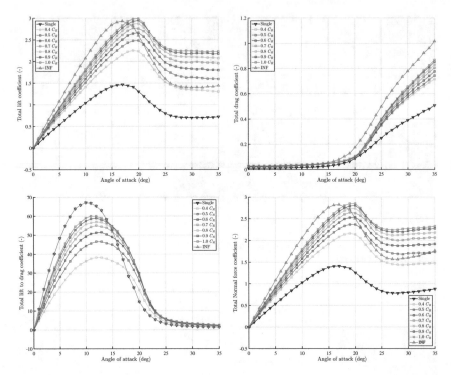

Fig. 4.9 Total hydrodynamic coefficients of a single NACA 0018 rudder and twin NACA 0018 rudders with various y_{TR}

The variation of y_{TR} has larger impacts on the lift coefficient of the port side rudder (C_L^{TP}) than the starboard side rudder (C_L^{TS}). An impacts in y_{TR} leads to an increase in C_L^{TP} and a relatively small decrease in C_L^{TS}. Smaller y_{TR} has larger impacts on the drag coefficient of the port side rudder (C_D^{TP}) and the drag coefficient of the starboard side rudder (C_D^{TS}), more specifically larger differences of C_D^{TP} and C_D^{TS} from the drag coefficient of the single rudder.

Figure 4.9 shows the total hydrodynamic coefficients of the twin rudders with different y_{TR}, where 'Single' indicates a single rudder while 'INF' means twin rudder with infinite y_{TR}. It is assumed that the hydrodynamic coefficients of 'INF' are two times those of 'Single'. The total lift coefficients of the twin rudders with larger y_{TR} are closer (before the stall angle) but further (after the stall angle) to the total lift coefficient of 'INF' than those of twin rudders with smaller y_{TR}. This phenomenon is caused by the stronger interaction that induces a more significant decrease in the lift coefficient on the port side when y_{TR} gets smaller as shown in Fig. 4.8.

The total lift coefficients and the total lift slopes increase with an increase in y_{TR} as the interaction effects decrease. An increase in y_{TR} slightly decreases the total drag coefficient at small angles of attack (smaller than 20°) while increases the total drag at large angles of attack. The total lift to drag ratios (C_L/C_D) decrease with a decrease of y_{TR} due to a larger decrease in lift and a smaller decrease in drag. Furthermore, twin rudders with large y_{TR} have higher normal force coefficients than those with small y_{TR}. On the whole, the two cases of the twin rudders with infinite y_{TR} (INF) and zero y_{TR} (Single) set the maximum and minimum bounds of the total twin-rudder hydrodynamic coefficients respectively. With impacts in y_{TR}, both the efficiency (the lift to drag ratio) and the effectiveness (the normal force) of the twin-rudder system in the open water increase.

4.2.4 Impacts of Profiles on Twin-Rudder Hydrodynamics

To analyze the impacts of the rudder profile on twin-rudder performance, this section presents the test results of 9 profiles from 3 families as listed in Table 4.1. Figures 4.10 and 4.11 present the pressure distributions and the stream traces of the test cases.

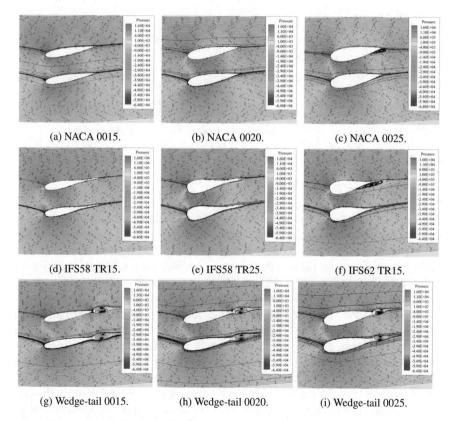

Fig. 4.10 Pressure distributions with stream traces of twin-rudder configurations with different profiles at an angle of attack of $10°$ with $y_{TR} = 0.5\ C_R$

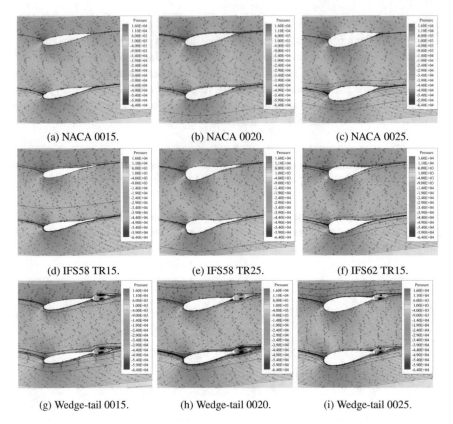

(a) NACA 0015.

(b) NACA 0020.

(c) NACA 0025.

(d) IFS58 TR15.

(e) IFS58 TR25.

(f) IFS62 TR15.

(g) Wedge-tail 0015.

(h) Wedge-tail 0020.

(i) Wedge-tail 0025.

Fig. 4.11 Pressure distributions with stream traces of twin-rudder configurations with different profiles at an angle of attack of $10°$ with $y_{TR} = 1.0\ C_R$

Thicker profiles have larger interaction effects as their wake field is larger than those of the thinner profiles. The non-streamlined tails recover the pressure at the trailing edge, increasing the lift but causing more drag. Furthermore, the wedge-tail profiles have larger recirculation than the well-streamlined profiles. In addition, the IFS profiles have a more significant decrease in the pressure at the leading edge, which partly explains their balanced performance in lift and drag.

The total lift, total drag, total lift to drag ratios, and total normal force of the tested profiles at y_{TR} of 0.5 and 1.0 C_R are presented in Figs. 4.12 and 4.13 respectively. These two values of spacing are the common minimum and maximum values for twin-rudder configurations. The impacts of y_{TR} on the hydrodynamic characteristics of different rudder profiles are similar, more specific impacts in y_{TR} increases the total lift coefficient, the slope of the total lift curves, and the total drag coefficient. Thus, the total normal force becomes larger, which enhances the effectiveness of the twin rudders. Since the increase in the total lift coefficient is larger than the

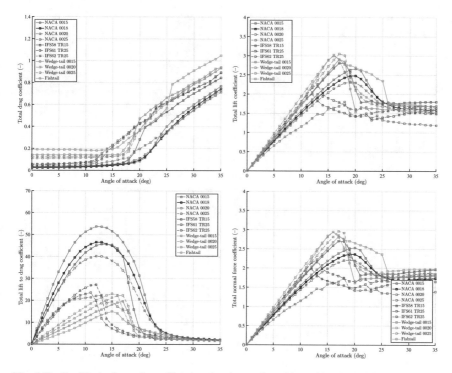

Fig. 4.12 Total hydrodynamic coefficients of various twin rudders with $y_{TR} = 0.5\ C_R$

impacts in the total drag coefficient, the total lift to drag ratio gets larger, leading to an improvement of the efficiency of the twin-rudder system.

Among the test profiles, in general, wedge-tail is most effective, NACA is most efficient, and IFS is balanced in efficiency and effectiveness. The thinner profiles have better performance than the thicker ones. Commonly, well-streamlined profiles (NACA and IFS) are better than those profiles with non-streamlined appendages (wedge-tail) in the efficiency but worse in the effectiveness. As a summary, the impacts of rudder profiles on the single-rudder and twin-rudder hydrodynamic coefficients are similar.

4.2.5 Impacts of Spacing on Quadruple-Rudder Hydrodynamics

To analyze the interaction effects between the two twin-rudder units in quadruple-rudder configurations, y_{TR} and y_{TU} are varied in 3 steps. In total, 9 combinations are tested and compared. The angles of attack are tested in a range of 0–15° to reduce the computation time. Figure 4.14 presents the pressure distributions and the stream

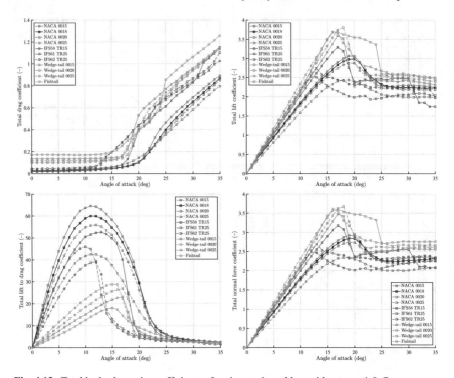

Fig. 4.13 Total hydrodynamic coefficients of various twin rudders with $y_{TR} = 1.0\ C_R$

traces of each case. As y_{TR} and y_{TU} increase, the pressure distribution of each rudder gets similar to that of the single rudder.

The lift and drag coefficients of each rudder in the quadruple-rudder configurations are compared to the single-rudder coefficients in Figs. 4.15 and 4.16. As shown in Fig. 4.4, the legends of Single, Average, QSO, QSI, QPI, and QPO stand for the single rudder, the average coefficient of the four rudders, the outer rudder of the starboard side twin-rudder unit, the inner rudder of the starboard side twin-rudder unit, the inner rudder of the port side twin-rudder unit, and the outer rudder of the port side twin-rudder unit respectively. Following the changes in the pressure distribution, the lift and drag coefficients of each rudder in quadruple-rudder configurations get closer to those of the single rudder as spacing among rudders increase.

The total hydrodynamic coefficients of the quadruple-rudder configurations are compared in Fig. 4.17. With the same y_{TR}, the quadruple-rudder configuration with a larger y_{TU} has higher total C_L and lower total C_D, thus higher total C_L/C_D and total C_N because the two inner rudders (QSP and QPS) are in the low-pressure field that is caused by the two outer rudders (QSS and QPP). When y_{TU} is the same, the lift coefficient increases and the drag coefficient decreases as y_{TR} increases. As a summary, the efficiency and the effectiveness of a quadruple-rudder configuration improve as y_{TR} and y_{TU} get larger.

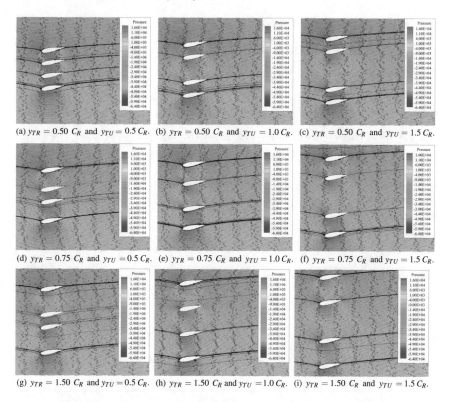

(a) $y_{TR} = 0.50\ C_R$ and $y_{TU} = 0.5\,C_R$. (b) $y_{TR} = 0.50\ C_R$ and $y_{TU} = 1.0\,C_R$. (c) $y_{TR} = 0.50\ C_R$ and $y_{TU} = 1.5\,C_R$.

(d) $y_{TR} = 0.75\ C_R$ and $y_{TU} = 0.5\,C_R$. (e) $y_{TR} = 0.75\ C_R$ and $y_{TU} = 1.0\,C_R$. (f) $y_{TR} = 0.75\ C_R$ and $y_{TU} = 1.5\,C_R$.

(g) $y_{TR} = 1.50\ C_R$ and $y_{TU} = 0.5\,C_R$. (h) $y_{TR} = 1.50\ C_R$ and $y_{TU} = 1.0\,C_R$. (i) $y_{TR} = 1.50\ C_R$ and $y_{TU} = 1.5\,C_R$.

Fig. 4.14 Pressure distributions with stream traces of quadruple-rudder configurations at an angle of attack of $10°$ with different y_{TR} and y_{TU}

4.3 3D RANS Study on Rudder Hydrodynamics

To analyze the effects of the shallow water and the effective aspect ratio on rudder hydrodynamics, this section presents the results of 3D RANS simulations with a NACA 0020 profile. The 3D RANS method is validated with wind tunnel results in Sect. 4.1.2. The test configuration is shown in Figure 3.9. Three geometric aspect ratios ($\Lambda_G = 1.2,\ 1.5,$ and 2.0) at two non-dimensional tip clearance ($d'_R = 0.2$ and 0.5) are tested. Furthermore, the validation case, where ($\Lambda_G = 1.5$ and $d'_R = 0.0$), is included to show the effect of d'_R on the effective aspect ratio. The lift coefficients, the drag coefficients, the lift to drag ratio, and the normal force coefficients of the test cases are presented in Fig. 4.18.

With the same non-dimensional tip clearance (d'_R), a decrease in the geometric aspect ratio (Λ_G) decreases the slopes of all the four coefficient curves, namely the lift

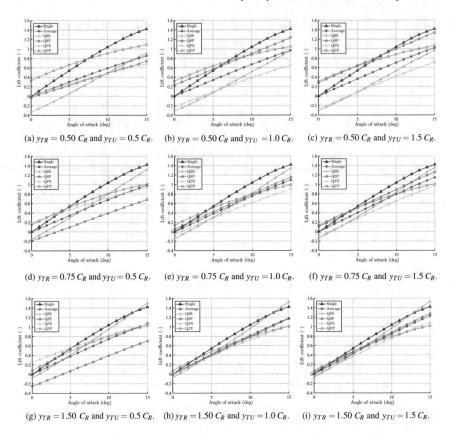

Fig. 4.15 Lift coefficients of each rudder in a quadruple-rudder configuration compared to the lift coefficient of a single rudder

coefficient, the drag coefficient, the lift to drag ratio, and the normal force coefficient while extends the stall angle. Therefore, a large Λ_G improves rudder efficiency and effectiveness as expected. Owing to the larger stall angles, the rudders with smaller Λ_G may be more effective at larger rudder angles than those with larger Λ_G. With the same Λ_G, the rudders with smaller d'_R have higher slopes of the four coefficient curves. As d'_R gets smaller, the shallow water effect appears, which enlarges the effective aspect ratio of the rudder.

It is known that the ratio of the effective and geometric aspect ratios (Λ_E / Λ_G) is 2 when the rudder tip connects with the bottom, which is the validation case for the 3D RANS method. Λ_E / Λ_G decreases with an increase in d'_R. As d'_R increases from 0.2 to 0.5, the reduction of Λ_E / Λ_G is more noticeable for the rudders with smaller Λ_G than

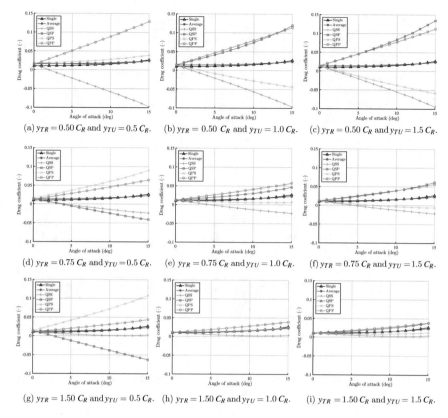

Fig. 4.16 Drag coefficients of each rudder in a quadruple-rudder configuration compared to the drag coefficient of a single rudder

those with larger Λ_G. This phenomenon shows that the shallow water (the small tip clearance) improves the efficiency and effectiveness of the rudder itself. Furthermore, it shows the benefit of using endplates to enhance the rudder performance, especially for rudders with small geometric aspect ratios.

4.4 Regression of Rudder Hydrodynamic Coefficients

Series of 2D RANS simulations of single-rudder, twin-rudder, and quadruple-rudder configurations are presented in Sect. 4.2. Based on these RANS results, rudder hydro-dynamic coefficients are obtained considering the rudder profile and the spacing

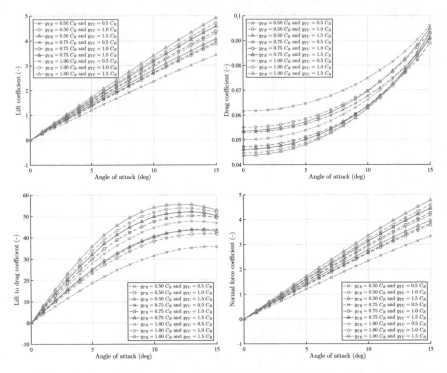

Fig. 4.17 Total hydrodynamic coefficients of quadruple NACA 0018 rudders with various y_{TR} and y_{TU}

among rudders. New regression formulas of the rudder hydrodynamic coefficients are presented in this section for the proposed maneuvering model in Chap. 6. The objective is to achieve C_L^{2D} and C_D^{2D} for Eq. 6.19 in Sect. 6.6.2.

4.4.1 Regression of Single-Rudder Hydrodynamic Coefficients

For a single 2D rudder in open water, Table 4.2 presents new regression formulas for various rudder profiles. These regression formulas are derived from the CFD results given in Fig. 4.6. The sample points are taken in the range of 0–10°. In this range, the accuracy of the RANS results is not significantly affected by the strong flow separation at large angles of attack.

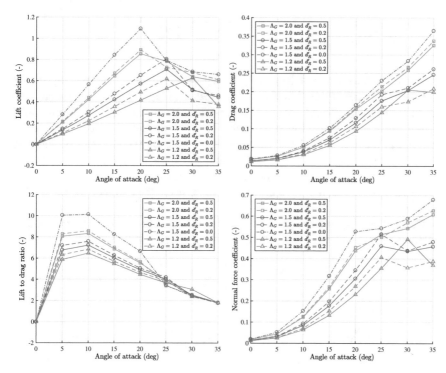

Fig. 4.18 Impacts of the shallow water and the effective aspect ratio on rudder hydrodynamic coefficients

Table 4.2 Regression coefficients for the rudder hydrodynamic characteristics of single rudders

Rudder profile	$\dfrac{\partial C_N^{2D}}{\partial \sin \alpha_R}$	$\dfrac{\partial C_L^{2D}}{\partial \sin \alpha_R}$	$C_{L_0}^{2D}$	$\dfrac{\partial C_D^{2D}}{\partial \sin \alpha_R}$	$C_{D_0}^{2D}$
NACA 0012	6.16	6.24	0.00	0.03	0.01
NACA 0015	6.10	6.18	0.00	0.03	0.01
NACA 0018	5.94	6.01	0.00	0.03	0.01
NACA 0020	5.78	5.85	0.00	0.03	0.01
NACA 0025	5.20	5.26	0.00	0.04	0.01
IFS58 TR15	6.78	6.85	0.00	0.03	0.01
IFS61 TR25	7.19	7.27	0.00	0.05	0.02
IFS62 TR25	7.02	7.10	0.00	0.08	0.01
Wedge-tail 0015	7.12	7.16	0.00	−0.01	0.07
Wedge-tail 0020	7.52	7.57	0.00	0.00	0.06
Wedge-tail 0025	7.94	8.00	0.00	0.01	0.05

4.4.2 Regression of Twin-Rudder Hydrodynamics Coefficients

In Sect. 4.2.3, the impacts of the spacing on twin-rudder hydrodynamic characteristics is discussed by modifying y_{TR}. Analyzing the data in Fig. 4.8, a correlation between the lift and drag coefficients of each rudder in twin-rudder configurations with those of the single-rudder case is found. Therefore, it is proposed to use the single-rudder lift (C_L^{2D}) and drag (C_D^{2D}) coefficients to predict the lift and drag coefficients for starboard and port side rudders (C_L^{TS}, C_L^{TP}, C_D^{TS}, and C_D^{TP}) in the twin-rudder configuration. In the regression study, the data points of twin-rudder configurations with spacing in the range of 0.5 and 1.0 C_R are selected as it is the commonly applied range in practice.

Considering the impacts of the rudder profile and the spacing on lift and drag coefficients, the regression formulas are made in the following format:

$$
\left.
\begin{aligned}
C_L^{TS,TP} &= \left(k_1^{C_L}\frac{C_R}{y_R^{TS,TP}}\left|\frac{C_R}{y_R^{TS,TP}}\right| + k_2^{C_L}\frac{C_R}{y_R^{TS,TP}} + 1\right)\frac{\partial C_L^{2D}}{\partial \alpha}\sin\left(k_3^{C_L}\frac{C_R}{y_R^{TS,TP}}\left|\frac{C_R}{y_R^{TS,TP}}\right| + k_4^{C_L}\frac{C_R}{y_R^{TS,TP}} + 1\right)\alpha \\
&\quad + k_5^{C_L}\frac{C_R}{y_R^{TS,TP}}\left|\frac{C_R}{y_R^{TS,TP}}\right| + C_{L_0}^{2D} \\
C_D^{TS,TP} &= \left(k_1^{C_D}\frac{C_R}{y_R^{TS,TP}}\left|\frac{C_R}{y_R^{TS,TP}}\right| + k_2^{C_D}\frac{C_R}{y_R^{TS,TP}} + 1\right)\frac{\partial C_D^{2D}}{\partial \alpha}\sin\left(k_3^{C_D}\frac{C_R}{y_R^{TS,TP}}\left|\frac{C_R}{y_R^{TS,TP}}\right| + k_4^{C_D}\frac{C_R}{y_R^{TS,TP}} + 1\right)\alpha \\
&\quad + k_5^{C_D}\frac{C_R}{y_R^{TS,TP}}\left|\frac{C_R}{y_R^{TS,TP}}\right| + C_{D_0}^{2D},
\end{aligned}
\right\} \quad (4.1)
$$

where

- $\frac{C_R}{y_R^{TS,TP}}$ concerns the relative position of each rudder in a twin-rudder configuration. y_R^{TS} and y_R^{TP} are the relative positions of the starboard side and the port side rudder to the propeller shaft.
- $\frac{\partial C_L^{2D}}{\partial \alpha}$ and $\frac{\partial C_D^{2D}}{\partial \alpha}$ are the slopes of the lift and drag coefficients of a single rudder, which are dependent on the rudder profile as shown in Table 4.2.
- $k_1^{C_L,C_D}\frac{C_R}{y_R^{TS,TP}}\left|\frac{C_R}{y_R^{TS,TP}}\right| + k_2^{C_L,C_D}\frac{C_R}{y_R^{TS,TP}} + 1$ considers the interaction effect on the lift and drag slopes.
- $k_3^{C_L,C_D}\frac{C_R}{y_R^{TS,TP}}\left|\frac{C_R}{y_R^{TS,TP}}\right| + k_4^{C_L,C_D}\frac{C_R}{y_R^{TS,TP}} + 1$ accounts for the interaction effect on the effective angle of attack.
- $k_5^{C_L,C_D}\frac{C_R}{y_R^{TS,TP}}\left|\frac{C_R}{y_R^{TS,TP}}\right|$ represents the interaction effect on the lift and drag coefficients at zero angle of attack.
- $C_{L_0}^{2D}$ and $C_{D_0}^{2D}$ are the lift and drag coefficients of a single rudder at zero angles of attack. $C_{L_0}^{2D}$ and $C_{D_0}^{2D}$ can be assumed as zero for symmetric well-streamlined profiles while they should be non-zero when asymmetric profiles are applied.

$\frac{\partial C_L^{2D}}{\partial \alpha}$ and $\frac{\partial C_D^{2D}}{\partial \alpha}$ are determined from the lift and drag coefficients of single-rudder at angles of attack in the range of $0°$ and $10°$. This range is determined to minimize the impacts of the strong flow separation around the stall angle on the RANS simulations.

Table 4.3 Regression coefficients of the twin-rudder lift and drag coefficients

Rudder profile	$k_1^{C_L}$	$k_2^{C_L}$	$k_3^{C_L}$	$k_4^{C_L}$	$k_5^{C_L}$	$R_{C_L}^2$	σ_{C_L}
NACA 0015	0.109	−0.493	−0.069	0.378	0.079	0.999	0.001
NACA 0018	−0.078	0.387	0.118	−0.514	0.067	0.999	0.010
NACA 0020	−0.064	0.372	0.098	−0.475	0.083	0.999	0.012
NACA 0025	−0.073	0.437	0.056	−0.401	0.064	0.993	0.024
IFS58 TR15	0.140	−0.572	−0.095	0.427	0.071	0.999	0.012
IFS61 TR25	0.176	−0.642	−0.076	0.389	0.169	0.999	0.017
IFS62 TR25	−0.729	1.101	0.710	−1.235	0.182	0.969	0.080
Wedge-tail 0015	0.156	−0.620	−0.092	0.437	0.003	0.998	0.015
Wedge-tail 0020	−0.084	0.415	0.167	−0.642	0.065	0.997	0.019
Wedge-tail 0025	−0.068	0.377	0.206	−0.713	0.168	0.996	0.030
Rudder profile	$k_1^{C_D}$	$k_2^{C_D}$	$k_3^{C_D}$	$k_4^{C_D}$	$k_5^{C_D}$	$R_{C_D}^2$	σ_{C_D}
NACA 0015	3.580	−1.524	−0.012	0.045	0.000	0.999	0.001
NACA 0018	2.729	−1.145	−0.002	0.019	0.000	0.999	0.001
NACA 0020	4.221	−1.562	−0.009	0.046	0.000	0.998	0.001
NACA 0025	4.506	−0.990	0.001	0.053	−0.001	0.996	0.002
IFS58 TR15	2.210	−1.360	−0.006	0.026	0.001	0.998	0.001
IFS61 TR25	3.444	−2.630	0.014	0.031	0.002	0.994	0.002
IFS62 TR25	1.749	−1.327	0.265	−0.400	0.003	0.993	0.002
Wedge-tail 0015	-7.663	−7.268	−0.208	0.512	0.001	0.979	0.002
Wedge-tail 0020	38.646	−5.670	−0.161	0.376	0.001	0.947	0.006
Wedge-tail 0025	17.057	−9.379	−0.113	0.274	0.001	0.993	0.003

The twin rudders are positioned at equal distances from the propeller shaft, therefore, $y_R^{TS} = \frac{1}{2}y_{TR}$ and $y_R^{TP} = -\frac{1}{2}y_{TR}$. The applied values of $\frac{\partial C_L^{2D}}{\partial \alpha}$, $\frac{\partial C_D^{2D}}{\partial \alpha}$, $C_{L_0}^{2D}$ and $C_{D_0}^{2D}$ for single profile are adapted from Table 4.2. Least squares fitting is performed in the format of Eq. 4.1 with the data presented in Fig. 4.4. The coefficients of the regression formulas for the lift and drag curves are presented in Table 4.3, where R^2 is R-squared (coefficient of determination) and σ is root mean squared error (standard error) of the regression coefficients.

The surface fitting and related residuals of the regression formulas with the coefficients in Table 4.3 are presented in Fig. 4.19 for the NACA 0018 profile as an example. The regression formulas are then used for different profiles at $y_{TR} = 0.5 \, C_R$ and $y_{TR} = 1.0 \, C_R$ and compared with the RANS results in Figs. 4.20, 4.21, 4.22, and 4.23 respectively. The comparison shows the usability of the equations for the test profile series.

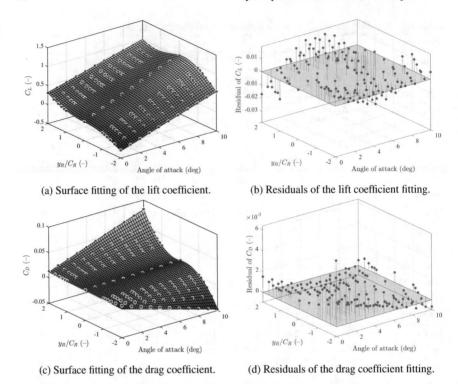

(a) Surface fitting of the lift coefficient.   (b) Residuals of the lift coefficient fitting.

(c) Surface fitting of the drag coefficient.   (d) Residuals of the drag coefficient fitting.

Fig. 4.19 Surface fitting of the lift and drag coefficients of the NACA 0018 profile with different spacing

4.4.3 Limitations of the Regression Formulas

In the previous sections, regression formulas have been proposed for single and twin rudders based on RANS simulations. These regression formulas are applied in Chap. 6 to build the integrated maneuvering model for inland vessels. They are also applicable to other research topics related to the rudder hydrodynamics with the following limitations: $|\alpha_R| <= 35°$ and $0.5\ C_R <= y_{TR} <= 1.0\ C_R$. For $|\alpha_R| > 35°$, the relative differences between the predicted value with the regression formula and the actual value would be enlarged. Normally, y_{TR} cannot be smaller than $0.5\ C_R$ from a practical point of view while the interaction effects get less significant when y_{TR} is larger than $1.0\ C_R$. Extra check on the usability of the proposed formulas is demanded when they are applied for cases that are out of the limitations.

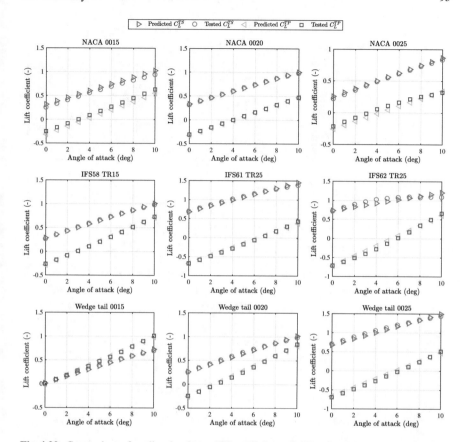

Fig. 4.20 Comparison of predicted and tested lift coefficients of different profiles in the twin-rudder configuration with $y_{TR} = 0.5\, C_R$

4.5 Summary

This chapter has introduced the applied 2D and 3D RANS methods, which are validated with experimental and numerical test results. The 2D RANS method is used to obtain hydrodynamic characteristics of rudders with infinite aspect ratios in open water. The impacts of the Reynolds numbers, the rudder profiles, and the spacing among multiple rudders on rudder hydrodynamics are analyzed. A 3D RANS method is utilized for the impacts of effective aspect ratios and shallow water on rudder hydrodynamics. Conclusions are summarized as follows:

- As the Reynolds number increases, the lift curve rises while the drag curve decreases. A Reynolds number of 6×10^6 can be considered as a threshold value above which little impacts on rudder hydrodynamics may be found.

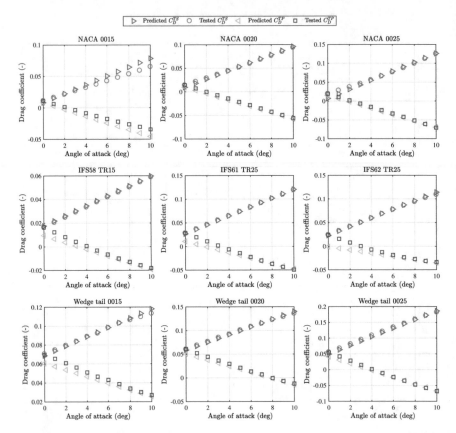

Fig. 4.21 Comparison of predicted and tested drag coefficients of different profiles in the twin-rudder configuration with $y_{TR} = 0.5\,C_R$

- Different profiles have different hydrodynamic characteristics. Among the tested profile families, the NACA series is most efficient while the wedge-tail series is most effective. Additionally, the IFS series is a good choice considering the balance of efficiency and effectiveness.
- An impact in y_{TR} increases the total lift coefficient, the total drag coefficient, the total lift to drag ratio, and the total normal force coefficient. Therefore, both the efficiency and the effectiveness of the twin-ruder system improve as the rudders are placed further apart.
- The profile does not significantly affect the impacts of y_{TR} on the twin-rudder hydrodynamics. The significance of the interaction effect is primarily determined by the spacing.
- The impacts of profiles on single-rudder and twin-rudder hydrodynamic coefficients are similar.

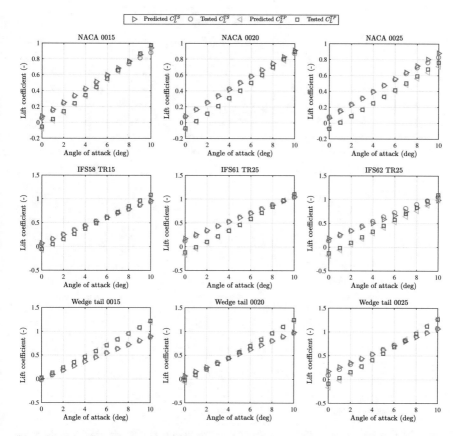

Fig. 4.22 Comparison of predicted and tested lift coefficients of different profiles in the twin-rudder configuration with $y_{TR} = 1.0\, C_R$

- As y_{TR} and y_{TU} increase, the total lift coefficient, the total lift to drag coefficient, and the total normal force coefficients of a quadruple-rudder configuration increase while the total drag coefficient decreases. Therefore, the efficiency and the effectiveness of the quadruple-rudder system improves as y_{TR} and y_{TU} get larger.
- In the same water depth, the lift coefficient, the drag coefficient, the lift to drag ratio, and the normal force coefficient increase with impacts in the geometric aspect ratio. A large geometric aspect ratio is recommended whenever it is applicable.
- Shallow water affects the effective aspect ratio of the rudder. A decrease in the rudder tip clearance increases the ratio of the effective geometric aspect ratio to the geometric aspect ratio. This increase is more significant for the rudders with small geometric aspect ratios. However, these shallow-water effects only occur in extraordinary shallow water.

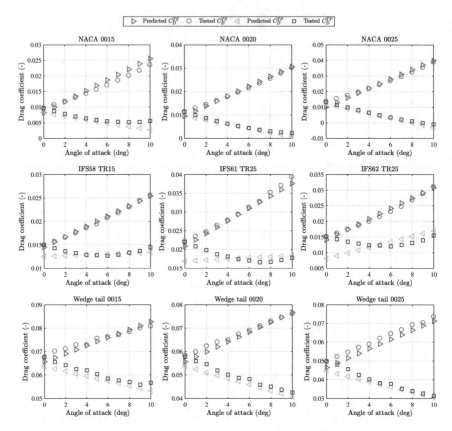

Fig. 4.23 Comparison of predicted and tested drag coefficients of different profiles in the twin-rudder configuration with $y_{TR} = 1.0\, C_R$

- It is recommended to use endplates to improve the rudder performance by increasing the effective aspect ratio, especially for the rudders with small aspect ratios. The impacts of the endplates on the rudder hydrodynamic coefficients for maneuvering will be discussed in Sect. 6.6.2.

With the results of this chapter, it is now possible to consider the impacts of the rudder configurations on ship maneuvering performance. Based on the RANS results that are obtained in this chapter, new regression formulas have been proposed and will be integrated into mathematical models for multiple-propeller multiple-rudder ships in Chap. 6. In Chap. 8, this mathematical model will be applied to carry out simulations of the reference ships in the proposed test maneuvers.

References

1. Ladson CL (1988) Effects of Independent Variation of Mach and Reynolds Numbers on the Low-Speed Aerodynamic Characteristics of the NACA 0012 Airfoil Section. Technical report, Langley Research Center, Hampton, Virginia, USA
2. Langley Research Center (2014) Turbulence Modeling Resource. Accessed on 10 Sept 2014, retrieved from http://turbmodels.larc.nasa.gov/naca0012_val.html
3. Thieme H (1965) Design of Ship Rudders (Zur Formgebung von Shiffsrudern). Technical report, Shipbuilding Institute, University of Hamburg, Washington, DC, USA
4. Molland AF, Turnock SR (1993b) Wind Tunnel Tests on the Influence of Propeller Loading on Ship Rudder Performance: Four Quadrant Operation, Low and Zero Speed Operation. Technical report, University of Southampton, Southampton, UK
5. Tang L (2008) Reynolds-Averaged Navier-Stokes simulation of low-Reynolds-number airfoil aerodynamics. J Aircraft 45(3):848–856
6. Catalano P, Tognaccini R (2011) RANS analysis of the low-Reynolds number flow around the SD7003 airfoil. Aerosp Sci Technol 15(8):615–626
7. Molland AF, Turnock SR (2007) Marine rudders and control surfaces: principles, data, design and applications, 1st edn. Elsevier Butterworth-Heinemann, Oxford

Chapter 5
A Primer of Inland Vessel Maneuverability

The maneuverability of seagoing ships has been studied extensively, but several factors make it doubtful to directly apply the seagoing ship oriented research to inland vessels. These factors relate to the environmental conditions in which inland vessels operate and the particulars of the vessels. Given the fact that seagoing ships and inland vessels are different, Sect. 5.1 describes the impact factors to be addressed for inland vessel maneuverability and identifies the rudder configuration as a crucial difference. Degrees of freedom and coordinate systems are described in Sect. 5.2 and Sect. 5.3 respectively. Kinematic and response motion models are presented and compared in Sect. 5.4. This monograph focuses on typical modular-type kinematic models of ship motions. Accordingly, Sect. 5.5 provides a link of the hydrodynamic forces with ship motions. Section 5.6 introduces the methods of maneuvering and resistance forces analysis through captive model tests, free-running tests, open water tests, and numerical simulations. Validation and verification of ship maneuvering modeling are discussed in Sect. 5.7. In the end, Sect. 5.8 concludes this chapter and presents a step-by-step procedure to study inland vessel maneuverability.

5.1 Impact Factors on Inland Vessel Maneuverability

Impact factors on inland vessel maneuverability are roughly characterized into two aspects: the external environment and the internal design. The external factors are set by the environment in which the ship sails while the internal factors are determined by the ship design and operation profiles. This section presents the literature on the impact factors and highlights the differences between inland vessels and seagoing ships. Sections 5.1.1 and 5.1.2 have addressed the additional impact factors on inland vessel maneuverability besides those for seagoing ships. Above all, existing estimation methods for hull forces are based on seagoing ships. They may not give proper results for inland vessels and result in a bad prediction of maneuverability. Therefore,

© Springer Nature Switzerland AG 2020
J. Liu, *Mathematical Modeling of Inland Vessel Maneuverability Considering Rudder Hydrodynamics*, https://doi.org/10.1007/978-3-030-47475-1_5

future research is suggested to consider the differences between seagoing ships and inland vessels while this monograph focuses on the impacts of rudder configurations on inland vessel maneuverability.

5.1.1   External Environment Factors

The following sections provide a review on two main impact factors of the navigation environment on the maneuverability of inland vessels, namely shallow-water and ship-ship/ship-bank interactions. External impacts are determined by the characteristics of the waterways while naval architects cannot control them.

5.1.1.1   Shallow-Water Effects

With the increasing demand for inland shipping capacity, inland vessels tend to be larger in size while depth and width of waterways remain the same. Meanwhile, large-scale sea-going ships get more common visit harbors with limited water depth. These practical issues lead to a concentrated focus on the shallow-water impacts as ship hydrodynamics and resultant motion characteristics may significantly change due to confined water depth. Any navigational or design analysis starts with an accurate determination of the hydrodynamic forces acted on the ship [1]. Scholars and researchers made a great effort to study the changes in wave patterns and resistance under shallow water conditions both theoretically and experimentally. Many empirical formulas and resistance charts have been proposed.

At first, the study focused on resistance neglected the viscosity of water and presume the water as a potential flow. In 1908, Havelock [2, 3] illustrated the main features of wave propagation in dispersive media produced by a point-pressure impulse traveling over a free surface, and in 1922 he developed the theory into shallow water condition. In 1953, Kinoshita and Inui [4] developed the interpolation formula of wave-making resistance raised by Hogner [5] into shallow water and they performed numerical calculations on two types of ship models which showed a good agreement with experimental results. Müller [6] came up with a formula for calculating the pressure field around the thin ship in finite water depth by arranging the sources and sinks on the longitudinal section.

Dawson [7] proposed the most influential non-linear method to study the wave produced by ships in 1977. Most of the current numerical methods of potential flow are developed from his method. In 1989, Yasukawa [8] raised a first-order panel method under shallow water by setting the Rankine source as the bottom boundary. In 1996, Kim [9] simulated the ship motion under finite water depth with linear and non-linear free surface boundary condition separately. Tarafder et al. [10, 11] used a second-order free surface condition to study wave-making resistance in shallow water with sinkage and trim effects. Tarafder [11] developed the method into a third order with the boundary elements.

As a consequence of scale enlargement in ships [12–14], the ship maneuvering performance in constrained channels is no longer only a concern for inland vessels but also crucial for seagoing ships. The performance of seagoing ships in shallow water is still considered to be less important because seagoing ships sail in deep water most of the time. On the contrary, for inland vessels which commonly sail in shallow water, shallow-water effects are very critical [15]. The ratio of the water depth to the ship draft (H/T) is typically larger than 4.0 for seagoing ships, but it is commonly smaller than 2.5 for inland vessels, especially in the dry period. The influence of water depth begins noticeable in medium-deep water $(1.5 < H/T < 3.0)$, becomes significant in shallow water $(1.2 < H/T < 1.5)$, and dominates in very shallow water $(H/T < 1.2)$ [16].

With the development of CFD technology, RANS methods have been applied to study shallow water problems considering water viscosity. Chen et al. [17] used two different RANSE solvers with structured and unstructured meshes, comparing with potential methods and EFD results. Jachowski [18] used the commercial solver ANSYS Fluent to simulate squat phenomena, which means the sinkage and trim by stern happened in shallow water. Raven [19] focused on the shallow water effects on viscous resistance by simulating double body viscous flow for 4 different ships. Haase et al. [20] calculated the shallow water resistance of a catamaran in both model-scale and full-scale. The results are compared with model tests and sea trials to study the scale effect in shallow water. Sadat-Hosseini et al. [21] used the URANS method to study the resistance performance of ship-ship interaction in shallow water.

Resistance estimation in shallow water plays an important role in ship design, especially for inland vessels. Several resistance charts have been developed to predict the shallow water resistance without detailed calculation for preliminary design. One classic method is raised by Schlichting [22]. Based on resistance value in deep water, he accounted for the influence of wave-making conditions and flow changes in shallow water with wave speed loss and re-flux speed to get the estimated resistance value in shallow water. Hofman and Kozarski [15] presented shallow water resistance charts obtained by systematic numerical calculation of Srettensky integral, for a variety of ship forms and dimension ratios. Also, the CFD methods are widely applied in resistance prediction. Peeters et al. [23] developed a CFD model to predict the resistance forces acting on inland vessels in shallow water. And a benchmark vessel for seagoing ships, the KVLCC2 hull, is taken to examine the effectiveness of this methodology.

The first experiment of ship resistance with the actual bottom effect was reported by Izubuchi and Nagasawa [24]. Previous Experiments on finite water depth problems were committed with false bottoms. There are many other towing tank tests among resent years, but shallow water towing tank tests were mostly carried out at relatively slow speed due to the unsteady effects. Doctors et al. [25] demonstrated that the unsteady effects are prominent and that it is essentially impossible to achieve a steady-state resistance curve in a towing tank of typical proportions for a water-depth-to-model-length ratio of 0.25. And proposed an application of unsteady linearized wave-resistance theory to present a simple application of the results to the planning and analysis of towing-tank tests. The additional hydrodynamic forces, squat and

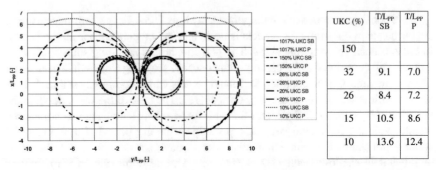

'SB'stands for the starboard side of the ship, while 'P' means the port side of the ship.

Fig. 5.1 Free-running turning circle model tests with a 6000 TEU container ship in deep and shallow water compared with the tactical diameter prediction (table) from captive model tests [34]

vertical motions of hulls sailing in shallow water are provided. Raven [26] addressed shallow-water ship model testing and the determination of water depth impacts on ship resistance. Also, a new procedure to correct ship speed trials for water-depth effects is introduced and validated with full-scale trial data.

Caplier et al. [27] reviewed the experimental methods developed in the towing tank of the Prime Institute of the University of Poitiers, France, for the characterization of ship wakes and drag forces in confined waters. The limited water depth influences ship resistance [28], which is the main component of the hull induced forces in the longitudinal direction (X_H). Meanwhile, in constrained waterways, the horizontal restrictions also influence the resistance, which is the so-called blockage effect [29, 30]. At the same advance speed, a decrease of H/T results in impacts in the ship squat and the wave height [18], leading to impacts in X_H. Gronarz [31, 32] proposed a shallow-water maneuvering model and indicated that the influence of shallow water on the hull dominates and that on the propeller and the rudder can be neglected. For inland vessels, Rotteveel [33] compared the existing models for ship resistance, proposed shallow water corrections for maneuvering, and pointed out that the current estimation methods do not represent the shallow-water effects in extremely shallow water.

Considering the lateral forces (Y_H) and moments (N_H), Eloot and Vantorre [34] reported an increase of course keeping ability leading to a larger tactical diameter in shallow water than in deep water for a slender seagoing container ship as shown in Fig. 5.1. International Towing Tank Conference [35] also presented an increase of the tactical diameters due to an increment of the hull damping force, which is a widely known shallow water effect. However, Yasukawa and Kobayashi [36, 37] discovered that the tactical diameter is smaller in shallow water than in deep water for twin-propeller wide-beam ships as shown in Fig. 5.2. This phenomenon is caused by the increase of rudder forces due to the high propeller load in shallow water.

Kijima and Nakiri [38], Lee and Lee [39] showed that shallow water effects on the tactical diameters and advances vary with ship profiles. As shown in Fig. 5.3, the

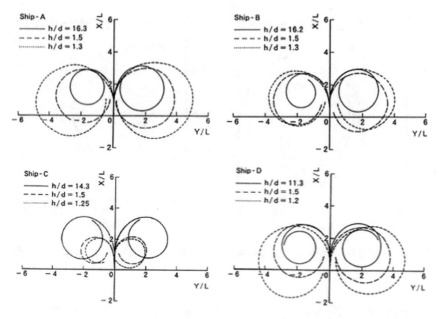

Fig. 5.2 Comparisons of ship trajectories in shallow water and deep water [37]

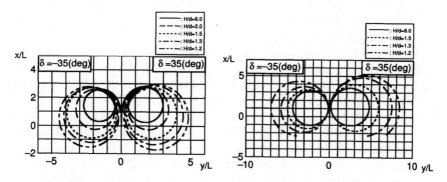

Fig. 5.3 Various shallow water effects on the tactical diameters and advances for different ship profiles [39]

tactical diameter increases while the advance almost remains the same on the left, but both the tactical diameter and the advance increase on the right. On the contrary to the above findings, Koh and Yasukawa [40] found that a pusher-barge system may have a smaller turning circle and a worse course keeping ability in shallow water than in deep water (Fig. 5.4).

Above all, research on shallow-water effects has been primarily done for large seagoing ships, maneuvering in ports or entering harbors, rather than for inland vessels [34, 36–39]. Shallow-water effects on the inland vessel maneuverability are still unclear. It should be noted that inland vessels have similar characteristics to the

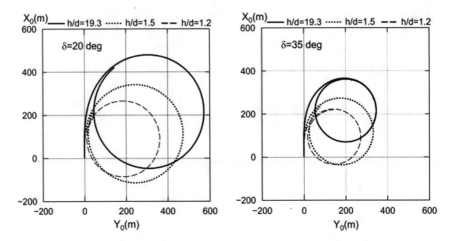

Fig. 5.4 Turning trajectories of a pusher-barge system at rudder angle 20° (left) and 35° (right) [40]

twin-propeller wide-width ship applied by Yasukawa and Kobayashi [36], Yoshimura and Sakurai [37], which means shallow-water effects on inland vessels could be different from typical large seagoing ships. To further study the ship maneuverability in shallow water, Landsburg et al. [14] suggested applying more accurate full-scale trials and mathematical modeling techniques.

Sano et al. [41] experimentally and numerically investigated the shallow water effect on the hydrodynamic force characteristics of two Wigley parabolic hulls moving side by side in close proximity with steering. Delefortrie et al. [42] presented the formulation of the 6-DOF maneuvering model in calm water for ships with conventional propulsion and steering as it is used on the ship maneuvering simulators at Flanders Hydraulics Research (FHR). The main input for this mathematical model is based on captive model tests. Furukawa et al. [43] presented shallow water effects on the longitudinal components of hydrodynamic derivatives based on the analysis of the measured forces in the database tested by the captive model tests. Van Zwijnsvoorde et al. [44] provided the benchmark data of captive and free running model tests with the DTC at full draft in calm water and in waves. Gourlay et al. [45] presented a comparison of calculated and measured vertical motions of the DTC container ship in shallow water. The considered conditions include calm water or head waves conditions at rest or with forward speed.

Yeo et a. [46] conducted experimental studies on the maneuvering characteristics of KVLCC 2 and numerical simulations of turning and zigzag maneuvers based on the hydrodynamic coefficients estimated from HPMM test results. Yeo et al. [47] provided the results from the benchmark HPMM tests, and the results of several supplemental tests carried out to check the overall reliability of the test results. Mei et al. [48] numerically and experimentally researched the wave-body interaction problems in finite water, including hydrodynamics forces, motion response and added resistance. The simulation results were compared with the results of FRMTs, especially focusing on the change in maneuverability due to change of water depth. Chen et al. [49] introduced a novel Nonlinear Least Squares–Support Vector Machine (NLS-SVM) method to accurately identify the ship's roll model parameters in shallow water, and solve the problems of difficult estimating nonlinear damping coefficients by traditional methods. Hanne and Bottner [50] provided the sea trials data for tuning and validation of maneuvering models, numerical as well as scaled physical model tests.

5.1.1.2 Ship-Ship Interactions

Inland waterways feature many artificial structures, such as locks, terminals, and bridge pillars along or in the channel, restricting the navigable area [51]. Thus, knowledge of ship behavior in horizontally and vertically restricted areas helps to reduce infrastructure and operation costs and enhance the navigation safety [34]. Inland vessels suffer far less from the strong wind and waves than seagoing ships. However, in natural waterways, currents may lead to very different sailing conditions for the upstream and downstream directions. In many cases, seagoing ships are requested to use additional maneuvering assistance like tugs in mandatory pilotage areas. On the contrary, inland vessels have to maneuver independently almost all the time. Consequently, inland vessels should meet higher maneuverability standards than seagoing ships to ensure safe navigation.

Due to the high density of traffic in inland waterways, inland vessels have to pass and meet each other at a close range much more frequently than seagoing ships. When a ship moves into the proximity of other ships, lateral force and yaw moment are induced due to the asymmetrical flow around the ship, which are the so-called ship-ship interactions. For seagoing ships, both numerical [52–54] and experimental [55–58] methods were applied to analyze these interactions and their impacts on ship maneuverability. However, for inland vessels, no systematic research was found.

5.1.1.3 Ship-Bank Interactions

Ship handling is also affected by the ship-bank interactions [59, 60]. The minimum channel width to the ship's width ratio may be as small as 4, 3, or 2 for double-lane, narrow-double lane, or single-lane channels respectively [61, 62]. Lee and Lee [63] researched the transverse distance and the maximum rudder angle for safe passing. Vantorre et al. [64] proposed an empirical formula to predict the ship-bank interaction forces through model tests. De Koning Gans [65], Lo [53], Zou and Larsson [66] investigated the sinkage and trim caused by ship-ship and ship-bank interactions through CFD simulations. Similar to the research on shallow-water effects, the scenarios for ship-ship and ship-bank interactions were primarily assumed for a large seagoing ship, for instance, the KVLCC2 tanker [66], in a narrow channel.

5.1.1.4 Concluding Remarks of External Environment Factors

In the external design factors, the shallow water has significant effects on hull forces, in particular for the resistance, but not much for the rudder forces. Since the shallow-water effects become significant when $H/T < 1.5$ and the presented reference vessels in this monograph (Table 6.1) are primarily designed for the deep downstream of the Rhine and the Yangtze River, the shallow-water effects on ship maneuverability are not explicitly discussed. Furthermore, there is no validation data publicly available for shallow water mathematical modeling of inland vessels. The ship-ship and ship-bank interactions are important for safe operations but not thoroughly studied as this monograph focuses on the design of individual inland vessels.

5.1.2 Internal Design Factors

Analyzing the navigation environment, naval architects may adapt ship particulars to compensate for the negative influences of the external disturbances. Four main features of inland vessels, which are different from seagoing ships, specifically slow speed, hull forms, propulsion, and rudder configurations, are addressed in the following paragraphs.

5.1.2.1 Slow Speed

The cruising speed of inland vessels (8–$28\,\mathrm{km\,h^{-1}}$) is often slower than that of seagoing ships (18–$36\,\mathrm{km\,h^{-1}}$). To compare the velocity among ships, the Froude number (Fr) is commonly used. Considering the range of ship dimensions and velocities, inland vessels commonly have smaller Fr than seagoing ships, leading to lower wake-making resistance. Moreover, slow speed also means a slow incidence velocity to the propeller and the rudder, affecting their performance. Even though slow-speed maneuvering is not the central concern for the design of most seagoing ships, their crucial impacts on safe operations deserve attention [67, 68].

Inland vessels that consistently sail at a slower speed than seagoing ships should be optimized to improve their slow-speed performance. ITTC Manoeuvring Committee [69] raised the necessity of standards for slow-speed maneuvering. Additionally, ITTC Manoeuvring Committee [70] presented an overview of the existing slow-speed maneuvering models for seagoing ships. Oh [71] compared the effectiveness of typical slow-speed maneuvering models. In sum, Eloot and Vantorre [72] concluded the opportunities and limitations of slow-speed models, emphasizing the differences between low-speed and ordinary-speed maneuvering.

5.1.2.2 Hull Forms

Inland vessels include motor vessels, pusher-barge systems, and towed-barge systems. Compared with the hull forms of seagoing ships, inland motor vessels in Europe commonly have larger block coefficients (C_b), much larger length to width ratios (L/B), and much larger width to draft ratios (B/T) due to the limits imposed on the draft (T), the length (L), and the width (B) of the vessel [73]. These differences in hull forms greatly influence the ship resistance and other hull-generated hydrodynamic forces and moments in maneuvering.

Pusher-barge and towed-barge systems are widely used in inland waterways all around the world. They are superior to motor vessels regarding transport capacity in shallow water. A significant amount of research has been conducted on the hydrodynamics of pusher-barge systems. Koh and Yasukawa [40], Koh et al. [75–77], Luo and Zhang [74], Maimun et al. [78] presented maneuvering research on pusher-barge systems through numerical or experimental methods. Tabaczek [79], Tabaczek et al. [80] analyzed the resistance and planar motions of a twin-screw inland vessel with different bow forms.

5.1.2.3 Propulsion

The propulsion of inland vessels is affected by the ship thrusters and appendages. Since propellers are not far from the free surface, inland motor vessels are commonly designed with a tunnel at the aft ship to improve the propeller inflow and prevent the propeller ventilation. In the process of model tests for ships with appendages, scale effects were found by Clement [81], Gregory and Beach [82] in determining the resistance. To relate test results to practical ships, Holtrop [83] covered the extrapolation methods of ships with multiple appendages and complex thrusters. As the propeller diameter of inland vessels is constrained by the water depth, multiple propellers, especially twin-propellers, are commonly installed. Kim et al. [84] reported worse turning but better course keeping and course changing abilities of a twin-propeller ship compared to a single-propeller ship at sea. For inland vessels, further research is necessary to determine the impacts of the propulsion on ship maneuverability.

5.1.2.4 Rudder Configurations

Compared to a common value of 2 for seagoing ships [85], the aspect ratio of inland vessel rudders is limited by the water depth to 1–2. Meanwhile, rudder orders for course corrections are more frequently called in inland waterways than at sea. To obtain sufficient maneuvering forces and moments, a configuration of multiple rudders per propeller is commonly used to increase the total rudder area. Additionally, inland vessels feature a larger range of rudder profiles, for instance, the Schilling rudder with additional endplates on the root and tip. The rudder profile largely affects the rudder hydrodynamic characteristics and has further impacts on maneuverability.

However, the impacts of the rudder design choices like profiles, parameters, types, and interactions among multiple rudders are not fully considered in the existing studies.

Nagarajan et al. [86] demonstrated a superior course keeping ability of a VLCC tanker with Schilling rudders to conventional Mariner rudders at constant engine torque under various encounter angles of the wind. Vantorre [87] carried out comprehensive model tests to determine the open water characteristics of several rudder configurations in a shallow water towing tank. Last but not least, inland vessels may use large rudder angles larger than 35°, which is the common maximum angle for seagoing ships, in hard maneuvering situations. When side movement is needed at slow speed, the rudder angles may reach 90°. Currently, most of the research on propeller and rudder performance has been done for deep water [73]. The necessity to adjust the existing propeller and rudder models for shallow water was proposed by Eloot and Vantorre [72].

5.1.2.5   Concluding Remarks of Internal Design Factors

In the internal design factors, slow-speed maneuvering is critical for hard maneuvering situations in port areas. The ranges of inland vessel particulars are different from those of seagoing ships. Thus, extra care should be given to the usability of seagoing oriented empirical formulas for hull forces and moments, which will be further discussed in Chap. 6. Propulsion impacts are relatively minor on the ship maneuverability as it is common in a steady state during maneuvering operations. Last but not least, the rudder configuration of a ship is important for its maneuvering performance. Furthermore, the rudder configuration is one of the most critical differences between inland vessels and seagoing ships, which has not been fully considered as yet.

5.2   Degrees of Freedom

A lot of effort has been put into the ship maneuvering models, but current work is primarily evolutionary rather than revolutionary. Ships are free to surge, sway, and heave in the direction of axes and roll, pitch, and yaw with respect to axes respectively as illustrated in Fig. 5.5. As it may affect the speed and accuracy in simulations, the required number of degrees of freedom should be seriously considered according to the navigation conditions and ship particulars. For inland vessels that rarely encounter severe waves and swell, a model with 3 DOF (Degree Of Freedom), including surge, sway, and yaw motions can give sufficient results. Considering the impacts of sinkage and trim, a 6 DOF model can be more suitable for ships in constrained waterways and shallow water. This monograph takes a 3 DOF model as the goal is to draw general conclusions about the rudder impacts on inland vessel maneuverability rather than study a specific ship in specific water depth.

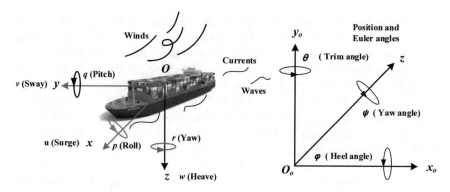
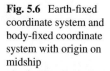

Fig. 5.5 Degrees of freedom of a ship

Fig. 5.6 Earth-fixed
coordinate system and
body-fixed coordinate
system with origin on
midship

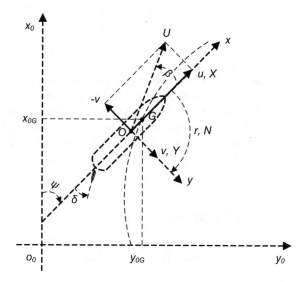

5.3 Coordinate Systems

To formulate the mathematical description of the ship motion, the coordinate system
should be established first. Figure 5.6 shows the coordinate systems used in the presented monograph. Two right-handed frames of axes are applied in a bidirectional
coordinate system: the earth-fixed coordinate system $o_0 - x_0y_0z_0$ and the body-fixed
coordinate system $o - xyz$. The origin of the earth-fixed coordinate system is commonly located at the start point of the maneuvering simulation, while the origin of
the body-fixed coordinate system is set on midship or at the center of gravity. As
highlighted by Yasukawa and Yoshimura [88], midship is more convenient than the
center of gravity considering ship load conditions.

x_0-axis is placed in the direction of the original course of the ship, while x-axis points towards the bow of the ship. The angle between x_0 and x is called the heading angle of ψ. y_0 and y axes are clockwise normal to the x_0 and x axes respectively. z_0 and z axes are defined downwards vertically. The ship's position is determined by the ship center of gravity in the earth-fixed coordinate x_{0G} and y_{0G} in the orientation of heading angle ψ. The center of gravity of the ship is situated at (x_G, y_G, z_G) in the body-fixed coordinate. x_G is the longitudinal coordinate of the center of gravity in $o - xyz$. y_G depends on the ship configuration, which is mostly zero but can be different, for instance in a push-barge system. z_G relates to the ship loading. With the origin of the body-fixed coordinate system lying at the center of gravity, the transformation from a body-fixed coordinate system to an earth-fixed coordinate system is shown as Eq. 5.1:

$$\left.\begin{aligned} x_0 &= x_{0G} + x \cos \psi - y \sin \psi \\ y_0 &= y_{0G} + x \sin \psi + y \cos \psi \\ z_0 &= z. \end{aligned}\right\} \tag{5.1}$$

Assuming the ship is maneuvering at contemporary speed U on midship, the speed is decomposed into an advance velocity u in the x-axis and a transverse v in the y-axis. The velocities at center of gravity can be expressed as $u_G = u$, $v_G = v + x_G r$. Drift angle β is the angle between the direction of actual velocity U and the x-axis achieving the relation $u = U \cos \beta$ and $v = -U \sin \beta$. Under the rudder angle δ, the ship is turning at an angular velocity $r = \dot{\psi}$ with respect to the midship.

5.4   Motion Models of Ships

5.4.1   Kinematic Motion Models

A lot of effort has been put into the ship maneuvering models, but current work is primarily evolutionary rather than revolutionary. Ships are free to surge, sway, heave in the direction of axes and roll, pitch, yaw with respect to axes respectively. As it may seriously affect the speed and accuracy in simulations, the required number of degrees of freedom should be seriously considered according to the navigation conditions and ship particulars. For inland vessels that rarely encounter severe waves and swell, a model with three degrees of freedom including surge, sway, and yaw motion may give sufficient results.

Extra dimensions should be added considering the ship particulars to improve the prediction accuracy. For example, it is critical to use a 4 DOF (degree of freedom) model including roll motion for ships with low metacentric height and high speeds. Stern and Agdrup [89] demonstrated the large difference in the prediction of maneuvering performance between 4 DOF and 3 DOF. Considering the impacts of sinkage and trim, a 6 DOF model can be more suitable for ships in constrained waterways and or in shallow water.

A system of non-linear three dimensional (surge, sway, and yaw) equations is discussed here as an example. The relation of kinematic motion and forces can be derived according to Newton's second law in the body-fixed coordinate system with the origin lying at the ship's center of gravity as Eq. 5.2:

$$\left.\begin{aligned} F_x &= m(\dot{u}_G - v_G r) \\ F_y &= m(\dot{v}_G + u_G r) \\ M_z &= I_{zG}\dot{r}, \end{aligned}\right\} \tag{5.2}$$

where F_x, F_y, M_z are external forces and moments acting at the center of gravity in the direction of the x-axis, y-axis, and around z-axis, m is the ship's mass, u_G, v_G are velocities at the center of gravity of the ship in the direction of the x-axis and y-axis respectively, r is the yaw rate, $\dot{u}_G$, $\dot{v}_G$, $\dot{r}$ are derivatives of the quantities, and I_{zG} is the moment of inertia of the ship around the center of gravity.

In practice, the origin of the body-fixed coordinate system is normally located on the midship point instead of at the center of gravity of the ship as shown in Fig. 5.6. Assuming the ship is symmetrical about its longitudinal plane, Eq. 5.2 can be adapted into Eq. 5.3 with the origin of the body-fixed coordinate system on midship rather than at the center of gravity,

$$\left.\begin{aligned} X &= m(\dot{u} - vr - x_G r^2) \\ Y &= m(\dot{v} + ur + x_G \dot{r}) \\ N &= I_z\dot{r} + mx_G(\dot{v} + ur), \end{aligned}\right\} \tag{5.3}$$

where X, Y, N are external forces and moment acting on midship in direction of the x-axis, y-axis, and around z-axis, u, v are velocities on midship in direction of the x-axis and y-axis respectively, and I_z is the moment of inertia of ship around midship.

5.4.2 Response Motion Models

To simplify the description of ship motions, Nomoto [90] derived a first-order response model from the equations of ship motion through the Laplace transformation as Eq. 5.4:

$$T\dot{r} + r = -K\delta, \tag{5.4}$$

where K is the turning ability index, T is the course keeping index. However, this simplified equation neglects the influences of transverse speed, longitudinal speed, and heel, resulting in an inaccurate prediction for most practical situations [91].

Norrbin [92, 93] introduced the non-linear damping factor into the response model as Eq. 5.5:

$$T\dot{r} + r + \alpha r^3 = -K\delta, \tag{5.5}$$

where α is a non-linear damping factor of the turning motions. However, this Norrbin model still does not include non-linear terms, for instance, the asymmetrical force induced by the single-screw propeller. Compared to the equations of ship motion, the response model is far less time-consuming. Therefore, response models are superior to study the course control problems or autopilot design [94, 95].

In this section, the coordinate system of maneuvering simulations is discussed. The coordinates are commonly used as a combination of the earth-fixed coordinate system and the body-fixed coordinate system. Attention should be paid to the origin of the body-fixed coordinate system and the directions of axes and angles for comparing prediction results from different research. Two maneuvering models, i.e. the equations of ship motion and the ship maneuvering response model, are compared to better express the external impacts. The equations of ship motion are superior in combining multiple components of forces and moments, which is thus proposed to be applied for inland shipping maneuverability prediction. The ship maneuvering response model is suggested to be used in autopilots because of the lower computation time.

5.5   Expressions of Hydrodynamic Forces

In order to solve the equations of motion (Eqs. 5.2 and 5.3) in Chap. 6, expressions are needed for the components of external forces (X and Y) and moment (N) acting on the ship. These forces and moments are caused by the ship hydrodynamics and environmental disturbances due to wind, waves, current, banks, and shallow water. Owing to the different manners of parameterization for the hydrodynamic forces and moments, the mathematical models are generally divided into two types: the whole ship model [96] and the modular model [92].

According to the typical maneuvers involved at each speed, Quadvlieg [73] proposed the requirements of mathematical models for speed in the range of low speed to service speed. Based on these requirements, Quadvlieg [73] indicated that a modular model is desired to describe the maneuvering behaviors of inland vessels. The modular model is superior in the physical representation of force components, such as hull, propeller, and rudder forces. These forces are functions of hydrodynamic coefficients which normally come from databases, captive model tests, approximation formulas, numerical methods, and system identification techniques. With the fast development of computer science, Computational Fluid Dynamics (CFD) techniques are increasingly popular in maneuverability research for hydrodynamic coefficients and force prediction.

Abkowitz [96] expressed the hydrodynamic forces and moments as functions of the kinematic parameters and the rudder angle as Eq. 5.6, which can be further expressed in Taylor-series as the following:

$$\left.\begin{array}{l} X = X(u, v, r, \dot{u}, \dot{v}, \dot{r}, \delta) \\ Y = Y(u, v, r, \dot{u}, \dot{v}, \dot{r}, \delta) \\ N = N(u, v, r, \dot{u}, \dot{v}, \dot{r}, \delta). \end{array}\right\} \tag{5.6}$$

Substituting Eq. 5.6 into Eq. 5.3, the equations for the whole ship (Abkowitz model) are achieved. This whole ship model treats the hull-water interaction as a black box and proved to be successful for arbitrary simulations [97, 98]. Represented by Ogawa and Kasai [99], the Mathematical Model Group of the Society of Naval Architects of Japan proposed a modular model, which is the so-called MMG model, like the following:

$$\left.\begin{array}{l} X = X_H + X_P + X_R \\ Y = Y_H + Y_P + Y_R \\ N = N_H + N_P + N_R. \end{array}\right\} \tag{5.7}$$

In the modular model, the hydrodynamic forces and moments are decomposed into three parts according to different origins, ship hull, propeller, and rudder, which are commonly denoted by the subscripts "H", "P", and "R". Consequently, the hull-propeller-rudder interactions and the performance of each component can be analyzed. To improve the accuracy and adaptability in different conditions, more followed work was done for ship maneuvering capacity in shallow water and constrained waterways. Fujino and Ishiguro [100] showed a remarkable dependence of the rudder effectiveness on the water depth. Kijima and Yasukawa [101, 102], Kijima et al. [103] proposed a modular prediction method and approximation formulas for hydrodynamic coefficients of hull forces in shallow and deep water. Kijima and Nakiri [101], Kijima and Qing [104], Kijima and Yasukawa [38] also applied the MMG model to analyze the interaction forces generated by other ships and banks.

Details about the MMG model were discussed in a series of papers written by Yasukawa and Yoshimura [105], Yoshimura [88, 106, 107]. Oltmann and Sharma [108] presented an approximation method of the interactions of hull-propeller-rudder in a combined engine and rudder model. Clarke [109] commented that the whole ship model gives a smooth representation of the forces, but has no physical meaning, while the second-order modulus expansions can well represent the hydrodynamic forces at angles of incidence. Comparing the advantages and disadvantages, the whole ship model is more suitable to obtain the overall performance of a maneuvering ship as a system through free-running tests or system identification methods, while Modulus expansions can better express the effects and interactions of each component, i.e. hull, propeller, and rudder.

5.6   Analysis of Hydrodynamic Forces

5.6.1   Maneuvering Force Analysis

To study the impacts of rudder configurations on ship maneuverability, the rudder induced forces and moments need to be analyzed. This section reviews the contemporary hydrodynamic force analysis methods to gather coefficients that are used to solve the equations of ship motions. These methods are captive model tests, free-running tests, open water tests, and numerical tests.

5.6.1.1   Captive Model Tests

Captive model tests are primarily used to determine the hydrodynamic coefficients of ships [110]. These tests are commonly carried out with Planar Motion Mechanism (PMM) or Computerized Planar Motion Carriage. The ship model (bare hull or equipped with a propeller, a rudder, and an electrical motor for propulsion) is attached to the towing carriage. Hydrodynamic coefficients are then obtained through testing in straight lines and harmonic tests. With the obtained hydrodynamic coefficients, other tests that are not suitable for the size of the towing basin, such as pull-out maneuvers and spiral tests, can be numerically simulated. Another kind of captive model tests, the so-called rotating-arm test, is designed to obtain stationary turning coefficients. The ship model is attached to a rotating arm which is set at the center of the basin turning at a constant velocity.

5.6.1.2   Free-Running Tests

Static towing tests are done at constant drift angles and rudder angles, which means that the motion is decoupled in the horizontal plane, so additional corrections have to be applied to the decoupling effects Van Leeuwen and Journée [110]. To directly obtain the maneuvering characteristics, free-running tests are applied to solve the uncoupled problems as it can generate a series of coefficients associated with positions Im and Seo [111]. International Towing Tank Conference [112] presented the standard procedure of free-running tests. System identification techniques are commonly correlated with the free-running model tests for hydrodynamic force coefficients [113–117]. Oltmann [118] gave an example of how to create a mathematical model from a series of zigzag tests. However, free-running tests are constrained by the dimension of the towing tank. Thus, these tests are regularly carried out on a rather small scale with high scaling errors.

5.6.1.3 Open Water Tests

In addition to tests for ship maneuverability, open water tests for propeller thrust, torque, and efficiency are usually performed in towing tanks or cavitation tunnels [112]. Furthermore, open water tests are also used to obtain rudder open water characteristics [87]. In contrary to the highly non-uniform ship wake in practice, these tests are carried out in uniform inflow. Considering the size of the towing tank, test models are commonly built in a range between 2 and 9 m in length [119]. Due to the differences of vortex shedding and flow separation, errors exist in scaling model test results to full scales [91]. Thus, Holtrop [83], Oyan [120] developed an extrapolation method to transfer model-scale results to full-scale applications [119].

5.6.1.4 Numerical Methods

Even though ship model tests are regarded as the most reliable solutions [121, 122], numerical methods are also widely applied. In practice, naval architects need simple prediction tools with reasonable accuracy. To design a new ship, coefficients may be obtained from one parent ship or a series of hull forms through empirical constants or regression formulas. Extensive tests are needed to build up these empirical databases. Longo et al. [125], Toda et al. [126], Todd [123, 124] described the procedure of tests on Series 60 and presented the results from different perspectives, such as wave profiles, wave elevations, mean velocities, pressure field distribution, and scale effects. As the computation power increases dramatically, numerical calculation through CFD methods is becoming more and more popular [127]. Chau [128] applied a Multi-block Finite Volume Method to study the free-stream characteristics of rudders.

El Moctar [129] applied RANS methods for rudders and propellers in uniform and oblique flow, rudder-propeller interaction, and forces on the ship hull in maneuvering. Morgan and Lin [130] gave an introduction to the historical development of the hydrodynamic prediction. Furthermore, it is possible to use CFD methods for model-scale or full-scale tests in virtual tanks [131–134]. CFD methods have been proved to be useful and more insights can be gained if additional work can be done on the verification and validation of the complex simulations as suggested by Rotteveel [33], Stern et al. [135, 136], Wilson et al. [137]. In order to compensate for the time-consuming CFD calculation and high-cost model tests, system identification methods were applied by Abkowitz [138], Araki et al. [116], Yoon and Rhee [114], Zhang and Zou [139].

5.6.1.5 Concluding Remarks of Maneuvering Force Analysis

After reviewing the existing methods, maneuverability prediction methods can be roughly divided into two aspects: free-running model tests and computer simulations based on mathematical models. Due to the high cost and uncontrollable test conditions, full-scale tests are rarely used in predictions. Hence, the model test is still considered as the most reliable measurement, which is commonly used as a final check before the ship construction. Computer simulations based on mathematical models are more flexible at the initial design stage. To obtain coefficients for the mathematical models, Sect. 5.6.1 presents literature on the contemporary hydrodynamic force analysis methods. The Computation Fluid Dynamics (CFD) method is selected because it does not require sophisticated physical test facilities. Furthermore, it can provide sufficiently accurate results at a relatively low cost.

5.6.2 Ship Resistance Analysis

It is of great necessity to predict the ship resistance in different sailing conditions at the first stage of ship design. This section concludes three common approaches to estimate the ship resistance. These three main ways are theoretical analyses, experimental tests, and numerical methods respectively.

5.6.2.1 Theoretical Analyses

The effective thrust should overcome the resistance so that the ship can sail. In general, actual ship resistance can be identified by tank tests or sea trials. In preliminary design stage, theoretical analyses are commonly used to approximate the ship resistance coefficient, especially some useful empirical formulas. Marchal et al. [140] evaluated the total resistance of an inland convoy with a polynomial formula and compared with three existing empirical methods. Tang and Han [141] put forward a formula for inland vessels analyzing the test data of barge, tugboat, passenger ships, and cargo ships in the middle and lower reaches of the Yangtze River.

Although using empirical formulas is simple and direct, it may lose accuracy when the target ship is out of the range of the test data sets. Also, some researchers used a variety of theories to analyzed the ship resistance. Han [142] analyzed the new slender-ship theory of wave resistance and proposed corresponding calculation methods. The theoretical calculation results agreed well with the well-found model experimental results. In consequence, the theoretical analyses are relatively low-cost, but the accuracy cannot be guaranteed without experimental results.

5.6.2.2 Experimental Tests

Experimental tests, which are divided into tank tests and sea trials, are used to get the actual ship resistance and regarded as the most reliable method. The tank tests are carried out in the towing tank under the condition of the same Froude number. The results need extrapolation to reach the real ship resistance but the sea trials can achieve directly. el Moctar et al. [143] aimed at a Post-Panamax container ship and proposed a case developed for benchmarking and validation of numerical methods. Kim et al. [144] tested three modern commercial hull forms and obtained the flow characteristics, which provides a good opportunity to explore integrated flow phenomena around practical hull forms of today. For inland waterway ships, Mucha et al. [145] proposed a novel test case for the provision of reference data for numerical methods.

5.6.2.3 Numerical Methods

Numerical methods have been developed rapidly for the past few years. These methods can use commercial codes to solve the governing equations, simulate and analyze the hydrodynamic problems, which are widely used in the shipping industry recently. To ensure the accuracy of numerical simulation, it is necessary to validate this method with experimental data first. After the validation, the simulations can be performed for different goals. Some researchers used the numerical methods to simulate the ship navigating at some given conditions. Demirel et al. [146] predicted the effect of biofouling on ship resistance using numerical methods. Guo et al. [147] combined the numerical methods and tank tests to explore the mechanism and correction method of the blockage effect. Kim et al. [148] performed numerical simulations for the prediction of added resistance and ship motions at various ship speed.

Other researchers put their attention on the numerical methods, trying to find out the key to precisely predict the ship resistance. Eça and Hoekstra [149] studied the accuracy of the numerical determination of the friction and pressure resistance coefficients of ship hulls for different levels of the iterative error and for computational domains of different size. Guo et al. [150] performed a systematic analysis of both experimental and numerical uncertainty and discussed the influence of turbulent models on the ship resistance prediction. Pereira et al. [151] focused on the calculation results between different RANS models. Islam et al. [152] studied the calm water resistance prediction of a container ship, then proposed a solver which could predict the resistance with reasonable accuracy utilizing minimum computational resources.

5.7 Validation and Verification

As more and more maneuvering capability predictions are done with computer modeling, the validation and verification for assessing the validity and quality of simulations are becoming more and more important. The subjective of verification is to ensure the model implementation and results are correct, while the goal of validation is to show that if the model is applicable under requirements with acceptable accuracy [153]. Sargent [154] presented the approaches to study the validity of simulation models, which were also adaptable for marine simulations. Stern et al. [136], Wilson et al. [137] presented the methodology and application of validation and verification of CFD simulations for container ships.

Maneuvering simulation models can be used for maneuverability prediction and ship maneuvering simulators [155]. Each component of the model simulation, such as ship particulars and environment data, mathematical modeling, hydrodynamic force prediction models, should be validated step by step. Comparable structures about how to document the parameters, methodology, and results were recommended by Endo [157], Hwang [156].

The validity and quality of the simulation results can be clearly presented by comparing it with full-scale test results. Trials of the tanker Esso Osaka had been widely cited since Crane [158] published the data. Simonsen and Stern [159, 160] validated a RANS model of rudder effects on maneuvering based on the Esso Osaka data. Other optional benchmark data for CFD validation were given by International Towing Tank Conference [161]. However, tests for Esso Osaka [158] were carried out on an old ship design using old equipment. Therefore, this set of data was not recommended by International Towing Tank Conference [155] as validation material.

As previously discussed by ITTC Manoeuvring Committee [162], full-scale benchmark data are expensive and it is hard to control the external disturbances. Thus, free running and captive model tests are favorable alternatives. Despite the scaling errors, model tests are superior in geometry control and test conditions. Four new benchmark models (two very large tankers KVLCC1 and KVLCC2, a container ship KCS, and a naval combatant DTMB 5415) were adopted by ITTC Manoeuvring Committee [163]. Though none of these models were built in full-scale, a series of model tests were done and compared by different institutions in the Workshop on Verification and Validation of Ship maneuvering Simulation Models [89, 164, 165]. These model parameters and test results are open to researchers and useful for various purposes of validation and verification.

Benchmark data mentioned above were applied to computational methods of seagoing ships in the deep open sea. Little information about experimental or numerical tests for inland vessels in deep or shallow water is publicly available. However, benchmark data for seagoing ships are still meaningful to develop a prototype maneuvering model for inland vessels. Benchmark data of model-scale and full-scale tests for inland vessels are proposed to be established for validation and verification of maneuverability prediction methods.

5.8   Summary

Ship maneuverability deserves more attention in ship design compared to the previous economic point of view [166], not only for seagoing ships but also for inland vessels. In order to identify and achieve a highly maneuverable ship, the main challenges are how to accurately calculate the forces and moments acting on the ship in different conditions, for instance, in shallow water. Moreover, the methods employed in the estimation process should be acceptable in time and cost while having adequate accuracy. Estimating the hydrodynamic forces on inland vessels with complex features is also a challenge. After all, with these challenges in mind, research should be carried out on standard maneuvers and criteria from the administration point of view and further give insight into the theoretical problems.

To obtain additional insight into inland vessel maneuverability evaluation and prediction methods, a primary of inland vessel maneuverability has been introduced with a step-by-step procedure, namely impact analysis, mathematical models, force expressions, maneuvering simulations, and validation and verification, as follows:

1. Regarding the main impact factors discussed in Sects. 5.1 and 5.1.2, collect data both on external conditions of the test and internal properties of the considered inland vessel as inputs for maneuvering simulations.
2. According to the estimated amplitudes of the ship motion in six dimensions, determine the number of degrees of freedom. Choose the position of the origin of the body-fixed coordinate system (at the center of gravity or on midship). Apply the equations of ship motion studied in Sect. 5.4.
3. Express the forces in the equations of motion to build up the mathematical model as discussed in Sect. 5.5.
4. Determine the hydrodynamic forces according to the mathematical model chosen or developed through model tests or numerical methods described in Sect. 5.6.
5. Based on the requirements in Chap. 7, evaluate the ship maneuvering performance.

Clear differences have been found in the external environment factors and internal design factors on ship maneuverability between inland vessels and seagoing ships. To have an accurate prediction of inland vessel maneuvering performance, estimation methods of forces and moments have to be adapted to the characteristics of inland waterways and inland vessels, such as shallow water conditions, ship-ship and ship-bank interactions, and multiple rudders per propeller configurations.

One of the most significant impacts is shallow water, which affects hull related forces (resistance, turning forces, and turning moments) and performance of propellers and rudders. However, research on shallow water effects was mainly done for seagoing ships in restricted waterways. Gaps need to be filled for inland vessels at slow speed in shallow water considering the complex configurations, such as the hull tunnel and multiple rudders.

Inland vessels commonly have complex navigation conditions and design features. As the modular model can better present various maneuvering related forces and moments including interactions among each component than the whole ship model, it is proposed to apply the modular model for inland vessel maneuverability prediction. While the whole ship model is suggested to be used with free-running tests for ship controller development. No benchmark data is publicly available for validation and verification materials for inland vessel maneuverability evaluation methods. Thus, benchmark test cases of model tests for inland vessels are suggested to be established.

The research on ship maneuverability is increasing, but still mainly focusing on seagoing ships rather than inland vessels. The main challenges for inland vessel maneuverability analysis are how to estimate the hydrodynamic forces with given ship particulars in a specified waterway and how to evaluate their performance according to maneuverability criteria required by the navigation environment of that waterway. These challenges will be further discussed in the following chapters.

References

1. Senthil Prakash MN, Chandra B (2013) Numerical estimation of shallow water resistance of a river-sea ship using cfd. Int J Comput Appl 71(5):33–40
2. Havelock TH (1908) The propagation of groups of waves in dispersive media, with application to waves on water produced by a travelling disturbance. Proc R Soc Lond Ser A, Contain Pap Math Phys Character 81(549):398–430
3. Havelock TH (1922) The effect of shallow water on wave resistance. Proc R Soc Lond Ser A, Contain Pap Math Phys Character 100(706):499–505
4. Kinoshita M, Inui T (1953) Wave-making resistance of a submerged spheroid, ellipsoid and a ship in a shallow sea. J Zosen Kiokai 75:119–135
5. Hogner E (1932) Eine interpolationsformel für den wellenwiderstand von schiffen. Jahrbuch der Schiffbautechnischen Gesellschaft 33:452–456
6. Müller E (1985) Analysis of the potential flow field and of ship resistance in water of finite depth. Int Shipbuild Prog 32(376):266–277
7. Dawson CW (1977) A practical computer method for solving ship-wave problems. In: Proceedings of second international conference on numerical ship hydrodynamics, pp 30–38
8. Yasukawa T (1989) Calculation of free-surface flow around a ship in shallow water by rankine source method. In: 5th international conference on numerical ship hydrodynamics, pp 643–653
9. Kim K (1996) Linear and nonlinear calculations of the free surface potential flow around ships in shallow water. In: 20th symposium on naval hydrodynamics, pp 408–425
10. Tarafder MS, Suzuki K, Kai H (2002) Wave making resistance of ships in shallow water based on second order free surface condition with sinkage and trim effects. J Kansai Soc Naval Archit Jpn 2002(237):2379–23,717
11. Tarafder S (2007) Third order contribution to the wave-making resistance of a ship at finite depth of water. Ocean Eng 34(1):32–44
12. Del Estado P (2000) Recommendations for the design of the maritime configuration of ports, approach channels and harbour basins. ROM 3.1-99
13. Gray WO, Waters J, Blume A, Landsburg AC (2003) Channel design and vessel maneuverability: next steps. Marine Technol 40(2):93–105
14. Landsburg AC, Barr RA, Daggett L, Hwang WY, Jakobsen B, Morris M, Vest L (2005) Critical needs for ship maneuverability: lessons from the Houston ship channel full-scale maneuvering trials. Marine Technol 42(1):11–20
15. Hofman M, Kozarski V (2000) Shallow water resistance charts for preliminary vessel design. Int Shipbuild Prog 47(449):61–76
16. Vantorre M (2003) Review of practical methods for assessing shallow and restricted water effects. International conference on marine simulation and ship maneuverability (MARSIM '03). Kanazawa, Japan, pp 1–11
17. Chen XN, Gronarz A, List S, Stuntz N (2001) Flow around ships sailing in shallow water experimental and numerical results. In: 23rd symposium on naval hydrodynamics
18. Jachowski J (2008) Assessment of ship squat in shallow water using CFD. Arch Civil Mech Eng 8(1):27–36
19. Raven HC (2012) A computational study of shallow-water effects on ship viscous resistance. In: 29th symposium on naval hydrodynamics, Gothenburg, Sweden

20. Haase M, Davidson G, Binns J, Thomas G, Bose N (2017) Full-scale resistance prediction in finite waters: a study using computational fluid dynamics simulations, model test experiments and sea trial measurements. Proc Inst Mech Eng Part M: J Eng Marit Environ 231(1):316–328

21. Sadat-Hosseini H, Wu PC, Toda Y, Carrica P, Stern F (2011) Urans studies of ship-ship interactions in shallow-water. In: 2nd international conference on ship manoeuvering in shallow and confined water: ship to ship interactions 2011, pp 299–308

22. Schlichting O (1934) Ship resistance in water of limited depth-resistance of sea-going vessels in shallow water. Jahrbuch der STG 35:127–148

23. Peeters G, Vanierschot M, et al (2019) Shallow water surge resistance identification for inland vessels. In: 5th international conference on ship manoeuvring in shallow and confined water, Flanders Hydraulics Research

24. Izubuchi T, Nagasawa S (1937) Experimental investigations on the influence of depth of water upon the resistance of ships. J Zosen Kiokai 61:165–206

25. Doctors LJ, Day AH, Clelland D (2008) Unsteady effects during resistance tests on a ship model in a towing tank. J Ship Res 52(4):263–273

26. Raven HC (2019) Shallow-water effects in ship model testing and at full scale. Ocean Eng 189(106):343

27. Caplier C, Gomit G, Rousseaux G, Calluaud D, Chatellier L, David L (2019) Calibrating and measuring wakes and drag forces of inland vessels in confined water in a towing tank. Ocean Eng 186(106):134

28. Tuck EO, Taylor PJ (1970) Shallow water problems in ship hydrodynamics. 8th symposium on naval hydrodynamics. USA, Washington, DC, pp 627–659

29. Kim HC, Moss JL (1963) Research in resistance and propulsion, Part III: blockage correction in a ship model towing tank and scale effect on propulsive parameters. Technical report, University of Michigan, Michigan, USA

30. Tamura K (1972) Study on the blockage correction. J Soc Naval Archit Jpn 17–28

31. Gronarz A (1993) A mathematical model for manoeuvring simulation on shallow water. International conference on marine simulation and ship maneuverability (MARSIM '93). St. John's, Newfoundland, Canada, pp 143–153

32. Gronarz A (1997) Numerical simulation of the ship's motion in manoeuvres with special consideration of the dependency of the water depth. PhD thesis, University of Duisburg

33. Rotteveel E (2013) Investigation of inland ship resistance, propulsion and manoeuvring using literature study and potential flow calculations. Master thesis, Delft University of Technology

34. Eloot K, Vantorre M (2011) Ship behaviour in shallow and confined water: An overview of hydrodynamic effects through EFD. In: Assessment of stability and control prediction methods for NATO air and sea vehicles, rTO-MP-AVT-189, pp 1–20

35. International Towing Tank Conference (2008) ITTC recommended procedures and guidelines 7.5-03-01: Uncertainty analysis in CFD verification and validation

36. Yoshimura Y, Sakurai H (1989) Mathematical model for the manoeuvring ship motion in shallow water (3rd report): manoeuvrability of a twin-propeller twin-rudder ship. J Kansai Soc Naval Archit Jpn 211:115–126

37. Yasukawa H, Kobayashi E (1995) Shallow water model experiments on ship turning performance. Mini symposium on ship manoeuvrability. Fukuoka, Japan, pp 72–83

38. Kijima K, Nakiri Y (2004) On the practical prediction method for ship manoeuvrability in restricted water (in Japanese). J Jpn Soc Naval Archit Ocean Eng 37–54

39. Lee CK, Lee SG (2005) On the ship manoeuvring characteristics in shallow water. Trans West-Jpn Soc Naval Archit 135–142

40. Koh KK, Yasukawa H (2012) Comparison study of a pusher-barge system in shallow water, medium shallow water and deep water conditions. Ocean Eng 46:9–17

41. Sano M, Yasukawa H, Kitagawa K, Yoshida S (2013) Shallow water effect on the hydrodynamic interaction between two ships with rudder in close proximity. 3rd international conference on ship manoeuvring in shallow and confined water. Ghent, Belgium, pp 113–121

42. Delefortrie G, Eloot K, Lataire E, Van Hoydonck W, Vantorre M (2016) Captive model tests based 6 DOF shallow water manoeuvring model. 4th international conference on ship manoeuvring in shallow and confined water. Hamburg, Germany, pp 273–286

43. Furukawa Y, Ibaragi H, Nakiri Y, Kijima K (2016) Shallow water effects on longitudinal components of hydrodynamic derivatives. In: 4th international conference on ship manoeuvring in shallow and confined water, pp 295–303
44. Van Zwijnsvoorde T, Delefortrie G, Lataire E (2019) Sailing in shallow water waves with the dtc container carrier: open model test data for validation purposes. In: 5th international conference on ship manoeuvring in shallow and confined water, Knowledge Centre Manoeuvring in shallow and confined water, pp 1–10
45. Gourlay T, Lataire E, Delefortrie G, Donatini L, Tello Ruiz M, Van Veen D, Bunnik T, Dallinga R (2019) Benchmarking of diffrac, fatima, hydrostar, moses, nemoh, octopus, pdstrip, rapid, seaway, slenderflow and wamit against measured vertical motions of the duisburg test case container ship in shallow water. In: 5th international conference on ship manoeuvring in shallow and confined water, Knowledge centre manoeuvring in shallow and confined water, pp 1–8
46. Yeo D, Yun K, Kim Y (2016) Experimental study on the manoeuvrability of kvlcc2 in shallow water. In: 4th international conference on ship manoeuvring in shallow and confined water, pp 287–294
47. Yeo D, Yun K, Kim Y, Kim S (2013) Benchmark hpmm tests for kcs in shallow water. In: 3rd international conference on ship manoeuvring in shallow and confined water: ship behaviour in locks, pp 249–255
48. Mei TL, Delefortrie G, Ruiz MT, Lataire E, Vantorre M, Chen C, Zou ZJ (2019) Numerical and experimental study on the wave-body interaction problem with the effects of forward speed and finite water depth in regular waves. Ocean Eng 192(106):366
49. Chen C, Ruiz MT, Delefortrie G, Mei T, Vantorre M, Lataire E (2019) Parameter estimation for a ship's roll response model in shallow water using an intelligent machine learning method. Ocean Eng 191(106):479
50. Hanne J, Bottner CU (2019) Sea trials for determination of manoeuvring charateristics in shallow water. In: 5th international conference on ship manoeuvring in shallow and confined water, Federal Waterways Engineering and Research Institute, pp 188–193
51. Liu J, Xu Y, Liu M (2012) Flow field simulation of bridge group area considering cross-bridge opening. In: International conference on marine simulation and ship maneuverability (MARSIM '12), Singapore
52. Chen HC, Lin WM, Hwang WY (2002) Application of Chimera RANS method for multiple-ship interactions in a navigation canal. In: 20th international offshore and polar engineering conference, Kitakyushu, Japan, vol 3, pp 330–337
53. Lo DC (2012) Numerical simulation of hydrodynamic interaction produced during the overtaking and the head-on encounter process of two ships. Eng Comput 29(1):83–101
54. Lima D, Sutulo S, Soares CG (2016) Study of ship-to-ship interaction in shallow water with account for squat phenomenon. In: 3rd international conference on maritime technology and engineering (MARTECH 2016), Lisbon, Portugal, pp 333–338
55. Vantorre M, Verzhbitskaya E, Laforce E (2002) Model test based formulations of ship-ship interaction forces. Ship Technol Res 49:124–141
56. Lataire E, Vantorre M, Delefortrie G (2009) Captive model testing for ship-to-ship operations. In: International conference on marine simulation and ship maneuverability (MARSIM '09), Panama City, Panama
57. Lataire E, Vantorre M, Vandenbroucke J, Eloot K (2011) Ship to ship interaction forces during lightering operations. 2nd international conference on ship manoeuvring in shallow and confined water: ship to ship interaction. Trondheim, Norway, pp 211–221
58. Eloot K, Delefortrie G, Vantorre M (2012) Inland navigation: Assessing the manoeuvring behaviour for real-time simulation purposes. In: International conference on marine simulation and ship maneuverability (MARSIM '12), Singapore, pp 1–12
59. Eloot K, Verwilligen J, Vantorre M (2007) A methodology for evaluating the controllability of a ship navigating in a restricted channel. Arch Civil Mech Eng 7(3):91–104
60. Lee CK, Lee SG (2008) Investigation of ship maneuvering with hydrodynamic effects between ship and bank. J Mech Sci Technol 22(6):1230–1236

61. Rijkswaterstaat (2011) Waterway Guidelines 2011
62. Rijkswaterstaat (2013) Supplement Rchtlijnen Vaarwegen 2011 (in Dutch)
63. Lee CK, Lee SG (2007) Hydrodynamic forces between vessels and safe maneuvering under wind-effect in confined waters. J Mech Sci Technol 21(5):837–843
64. Vantorre M, Delefortrie G, Eloot K, Laforce E (2003) Experimental investigation of ship-bank interaction forces. International conference on marine simulation and ship maneuverability (MARSIM '03). Kanazawa, Japan, pp 1–9
65. De Koning Gans H (2005) Squat results from calculations of panel methods. In: International maritime-port technology and development conference, Singapore
66. Zou L, Larsson L (2013) Computational Fluid Dynamics (CFD) prediction of bank effects including verification and validation. J Marine Sci Technol 18(3):310–323
67. Hwang W, Jakobsen B, Barr R, Ankudinov V, Fuller N, Vest L, Morris M, McGovern A, Landsburg A (2003) An exploratory study to characterize ship manoeuvring performance at slow speed. In: International conference on marine simulation and ship maneuverability (MARSIM '03), vol 3, Kanazawa, Japan
68. Dand I (2003) Low-speed manoeuvring criteria: some considerations. In: International conference on marine simulation and ship maneuverability (MARSIM '03), Kanazawa, Japan
69. ITTC Manoeuvring Committee (2008) Final report and recommendations to the 25th ITTC. In: 25th International Towing Tank Conference, Fukuoka, Japan
70. ITTC Manoeuvring Committee (2011) Final report and recommendations to the 26th ITTC. In: 26th International Towing Tank Conference, Rio de Janeiro, Brazil, vol 1, pp 123–181
71. Oh HW (2012) Advanced fluid dynamics. InTech
72. Eloot K, Vantorre M (2004) Prediction of low speed manoeuvring based on captive model tests: opportunities and limitations. International marine simulation forum (IMSF): annual general meeting 2004. Antwerp, Belgium, pp 13–17
73. Quadvlieg F (2013) Mathematical models for the prediction of manoeuvres of inland ships: does the ship fit in the river? In: Rigo P, Wolters M (eds) Smart rivers 2013. PIANC, Liège, Belgium/Maastricht, The Netherlands, pp 1871–1879
74. Luo W, Zhang P (2007) Evaluating the maneuverability of a new type of self-propelled barge. J Marine Sci Appl 6(4):44–47
75. Koh KK, Yasukawa H, Hirata N (2008) Hydrodynamic derivatives investigation of unconventionally arranged pusher-barge systems. J Marine Sci Technol 13(3):256–268
76. Koh KK, Yasukawa H, Hirata N (2008) Shallow water effect on turning motion of a pusher-barge system. In: 4th Asia-pacific workshop on marine hydrodynamics, vol 3 Taipei, pp 16–18
77. Koh KK, Yasukawa H, Hirata N, Kose K (2008) Maneuvering simulations of pusher-barge systems. J Marine Sci Technol 13(2):117–126
78. Maimun A, Priyanto A, Muhammad AH, Scully CC, Awal ZI (2011) Manoeuvring prediction of pusher barge in deep and shallow water. Ocean Eng 38:1291–1299
79. Tabaczek T, Kulczyk J, Zawiślak M (2007) Analysis of hull resistance of pushed barges in shallow water. Pol Marit Res 14:10–15
80. Tabaczek T (2010) Numerical simulation of planar motion of a twin-screw inland waterway vessel in shallow water. 18th international conference on hydrodynamics in ship design. Safety and operation, Gdansk, Poland, pp 37–50
81. Clement EP (1957) Scale effect on the drag of a typical set of planning boat appendages. Technical report, David Taylor Model Basin, Bethesda, Maryland, USA
82. Gregory D, Beach T (1979) Resistance measurements of typical planning boat appendages. Technical report, David Taylor Model Basin, Bethesda, Maryland, USA
83. Holtrop J (2001) Extrapolation of propulsion tests for ships with appendages and complex propulsors. Marine Technol 38(3):145–157
84. Kim YG, Kim SY, Kim HT, Lee SW, Yu BS (2007) Prediction of the maneuverability of a large container ship with twin propellers and twin rudders. J Marine Sci Technol 12(3):130–138
85. Kim HJ, Kim SH, Oh JK, Seo DW (2012) A proposal on standard rudder device design procedure by investigation of rudder design process at major Korean shipyards. J Marine Sci Technol 20(4):450–458

86. Nagarajan V, Kang DH, Hasegawa K, Nabeshima K (2008) Comparison of the mariner Schilling rudder and the mariner rudder for VLCCs in strong winds. J Marine Sci Technol 13:24–39
87. Vantorre M (2001) Stationary and non-stationary open water rudder tests. In: Kijima K (ed) Mini symposium on prediction of ship manoeuvring performance. Tokyo, Japan, pp 103–111
88. Yasukawa H, Yoshimura Y (2015) Introduction of MMG standard method for ship maneuvering predictions. J Marine Sci Technol 20(1):37–52. https://doi.org/10.1007/s00773-014-0293-y
89. Stern F, Agdrup K, (2008) SIMMAN 2008 workshop on verification and validation of ship maneuvering simulation methods. In: SIMMAN workshop, Denmark, Copenhagen
90. Nomoto K (1957) On the steering qualities of ships. Int Shipbuild Prog 4(35):
91. Bertram V (2012) Practical ship hydrodynamics, 2nd edn. Elsevier, Oxford
92. Norrbin NH (1970) Theory and observation on the use of a mathematical model for ship maneuvering in deep and confined waters. 8th symposium on naval hydrodynamics. Pasadena, California, pp 807–904
93. Norrbin NH (1978) A method for the prediction of the manoeuvring lane of a ship in a channel of varying width. In: Symposium on aspects of navigability of constraint waterways, including harbour entrances, vol 3 Delft, The Netherlands
94. Van Amerongen J, Udink Ten Cate AJ (1975) Model reference adaptive autopilots for ships. Automatica 11:441–449
95. Van Amerongen J (1984) Adaptive steering of ships: a model reference approach. Automatica 20(1):3–14
96. Abkowitz MA (1964) Lectures on Ship Hydrodynamics Steering and Maneuverability. Technical report, Hydro and Aerodynamic Laboratory, Lyngby, Denmark
97. Strom Tejsen J, Chislett MS (1966) A model testing technique and method of analysis for the prediction of steering and manoeuvring qualities of surface vessels. Technical report, Hydro and Aerodynamics Lab, Washington, DC, USA
98. Crane CL (1973) Maneuvering safety of large tankers: stopping, turning and speed selection. SNAME Trans 81:218–242
99. Ogawa A, Kasai H (1978) On the mathematical model of manoeuvring motion of ships. Int Shipbuild Prog 25(292):306–319
100. Fujino M, Ishiguro T (1984) A study of the mathematical model describing manoeuvring motions in shallow water. J Soc Naval Archit Jpn 156:180–192
101. Kijima K, Yasukawa H (1985) Manoeuvrability of ships in narrow waterway. J Soc Naval Archit Jpn 23(3):25–37
102. Kijima K, Katsuno T, Nakiri Y, Furukawa Y (1990) On the manoeuvring performance of a ship with the parameter of loading condition. J Soc Naval Archit Jpn 168:141–148
103. Kijima K, Tanaka S, Matsunga M, Hori T (1992) Manoeuvring characteristics of a ship in deep and shallow waters as a function of loading condition. 2nd International conference on manoeuvring and control of marine craft. Southampton, UK, pp 73–86
104. Kijima K, Qing H (1988) Manoeuvring motion of a ship in the proximity of bank wall. J Soc Naval Archit Jpn 162:125–132
105. Yoshimura Y (1986) Mathematical model for the manoeuvring ship motion in shallow water (in Japanese). J Kansai Soc Naval Archit Jpn 41–51
106. Yoshimura Y (1988) Mathematical model for the manoeuvring ship motion in shallow water (2nd report): mathematical model at slow forward speed. J Jpn Soc Naval Archit Ocean Eng 77–84
107. Yoshimura Y (2005) Mathematical model for manoeuvring ship motion (MMG Model). Workshop on mathematical models for operations involving ship-ship interaction. Tokyo, Japan, pp 1–6
108. Oltmann P, Sharma SD (1984) Simulation of combined engine and rudder maneuvers using an improved model of hull-propeller-rudder interactions. 15th symposium on naval hydrodynamics. Hamburg, Germany, pp 1–24

109. Clarke D (2003) The foundations of steering and manoeuvring. In: IFAC conference on manoeuvring and control of marine craft (MCMC 2003), Girona, Spain
110. Van Leeuwen G, Journée JMJ (1970) Prediction of ship manoeuvrability making use of model tests. Technical report, Ship Hydromechanics Laboratory, Delft University of Technology, Delft, The Netherlands
111. Im NK, Seo JH (2010) Ship manoeuvring performance experiments using a free running model ship. Int J Marine Navig Saf Sea Transp 4(1):29–33
112. International Towing Tank Conference (2008) ITTC Recommended procedures and guidelines 7.5-02-03-02.1: Testing and extrapolation methods propulsion, propulsor open water test
113. Grochowalski S (1989) Investigation into the physics of ship capsizing by combined captive and free-running model tests. John's, Newfoundland, Canada, Technical report, National Research Council Canada, St
114. Yoon HK, Rhee KP (2003) Identification of hydrodynamic coefficients in ship maneuvering equations of motion by estimation-before-modeling technique. Ocean Eng 30:2379–2404
115. Yoon HK, Son NS, Lee GJ (2007) Estimation of the roll hydrodynamic moment model of a ship by using the system identification method and the free running model test. IEEE J Ocean Eng 32(4):798–806
116. Araki M, Sadat-Hosseini H, Sanada Y, Tanimoto K, Umeda N, Stern F (2012) Estimating maneuvering coefficients using system identification methods with experimental, system-based, and CFD free-running trial data. Ocean Eng 51:63–84
117. Bonci M, Viviani M, Broglia R, Dubbioso G (2015) Method for estimating parameters of practical ship manoeuvring models based on the combination of RANSE computations and System Identification. Appl Ocean Res 52:274–294
118. Oltmann P (1996) On the influence of speed on the manoeuvring behaviour of a container carrier. International conference on marine simulation and ship maneuverability (MARSIM '96). Denmark, Copenhagen, pp 515–523
119. International Towing Tank Conference (2011) ITTC Recommended procedures and guidelines 7.5-01-01-01: Ship Models
120. Oyan E (2012) Speed and powering prediction for ships based on model testing. Master thesis, Norwegian University of Science and Technology
121. Vantorre M (1992) Accuracy and optimization of captive ship model tests. In: PRADS '92: 5th international symposium on practical design of ships and mobile units, Newcastle, UK, vol 1, pp 190–203
122. Vantorre M, Eloot K (1998) Requirements for standard harmonic captive manoeuvring tests. Control Eng Pract 6:643–652
123. Todd FH (1953) Some further experiments on single screw merchant ship forms - Series 60. SNAME Trans 61
124. Todd FH (1963) Series 60 methodical experiments with models of single-screw merchant ships. Technical report, David Taylor Model Basin, Washington, USA
125. Longo J, Stern F, Toda Y (1993) Mean-flow measurements in the boundary layer and wake and wave field of a Series 60 CB = 0.6 ship model, Part 2: scale effects on near-field wave patterns and comparisons with inviscid theory. J Ship Res 37(1):16–24
126. Toda Y, Stern F, Longo J (1992) Mean-flow measurements in the boundary layer and wake and wave field of a Series 60 CB = 0.6 ship model, Part 1: froude numbers 0.16 and 0.316. J Ship Res 36(4):360–378
127. Tyagi A, Sen D (2006) Calculation of transverse hydrodynamic coefficients using computational fluid dynamic approach. Ocean Eng 33:798–809
128. Chau SW (1997) Numerical investigation of free-stream rudder characteristics using a multi-block finite volume method. PhD thesis, University of Hamburg
129. El Moctar O (2001) Numerical computations of flow forces in ship manoeuvring. Ship Technol Res 48(3):98–123
130. Morgan WB, Lin WC (1998) Predicting ship hydrodynamic performance in today's world. Naval Eng J 110:91–98

131. Badoe CE, Phillips AB, Turnock SR (2015) Influence of drift angle on the computation of hull-propeller-rudder interaction. Ocean Eng 103:64–77
132. Kim H, Akimoto H, Islam H (2015) Estimation of the hydrodynamic derivatives by RANS simulation of planar motion mechanism test. Ocean Eng 108:129–139
133. Tezdogan T, Demirel YK, Kellett P, Khorasanchi M, Incecik A, Turan O (2015) Full-scale unsteady RANS CFD simulations of ship behaviour and performance in head seas due to slow steaming. Ocean Eng 97:186–206
134. Carrica PM, Mofidi A, Eloot K, Delefortrie G (2016) Direct simulation and experimental study of zigzag maneuver of KCS in shallow water. Ocean Eng 112:117–133
135. Stern F, Wilson RV, Coleman HW, Paterson EG (1999) Verification and validation of CFD simulations. Technical report, Iowa Institute of Hydraulic Research, Iowa City, USA
136. Stern F, Wilson RV, Coleman HW, Paterson EG (2001) Comprehensive approach to verification and validation of CFD simulations, Part 1: methodology and procedures. J Fluids Eng 123:793–802
137. Wilson RV, Stern F, Coleman HW, Paterson EG (2001) Comprehensive approach to verification and validation of CFD simulations, Part 2: Application for RANS simulation of a cargo/container ship. J Fluids Eng 123:803
138. Abkowitz MA (1980) Measurement of hydrodynamic characteristics from ship maneuvering trials by system identification. SNAME Trans 88:283–318
139. Zhang XG, Zou ZJ (2011) Identification of Abkowitz model for ship manoeuvring motion using ϵ-support vector regression. J Hydrodyn 23(3):353–360
140. Marchal JLJ, Shen YD, Kicheva D (1995) An empirical formula to estimate the resistance of a convoy in a restricted waterway. Marine Technol Sname News 40(1498):107–112
141. Tang T, Han H (1992) Resistance calculation formula for inland river transport vessels. Ship Eng 21–23
142. Han DF (2003) Research on the method and calculation of the new slender-ship wave resistance theory. J Hydrodyn(SerB) 15(4):124–124
143. el Moctar O, Shigunov V, Zorn T (2012) Duisburg test case: post-panamax container ship for benchmarking. Ship Technol Res 59(3):50–64. https://doi.org/10.1179/str.2012.59.3.004
144. Kim WJ, Van SH, Kim DH (2001) Measurement of flows around modern commercial ship models. Exp Fluids 31(5):567–578
145. Mucha P, Moctar OE, Dettmann T, Tenzer M (2017) Inland waterway ship test case for resistance and propulsion prediction in shallow water. Ship Technol Res 64(466):1–8
146. Demirel YK, Turan O, Incecik A (2017) Predicting the effect of biofouling on ship resistance using cfd. Appl Ocean Res 62:100–118
147. Guo C, Xu P, Wang C, Kan Z (2019) Numerical and experimental study of blockage effect correction method in towing tank. China Ocean Eng 33(5):522–536
148. Kim M, Turan O, Day S, Incecik A (2018) Numerical studies on added resistance and ship motions of kvlcc2 in waves. In: Trends and challenges in maritime energy management. Springer, pp 111–126
149. Eça L, Hoekstra M (2009) On the numerical accuracy of the prediction of resistance coefficients in ship stern flow calculations. J Marine Sci Technol 14(1):2–18
150. Guo BJ, Deng GB, Steen S (2013) Verification and validation of numerical calculation of ship resistance and flow field of a large tanker. Ships Offshore Struct 8(1):3–14
151. Pereira FS, Eça L, Vaz G (2017) Verification and validation exercises for the flow around the { KVLCC2 } tanker at model and full-scale reynolds numbers. Ocean Eng 129:133–148
152. Islam H, Rahaman MM, Akimoto H, Islam MR (2017) Calm water resistance prediction of a container ship using reynolds averaged navier-stokes based solver. Proc Eng 194:25–30
153. Garret M (1974) Statistical validation of simulation models. In: 11th conference on Winter simulation (WSC '79), San Diego, California, USA, vol 2, pp 497–503
154. Sargent RG (2005) Verification and validation of simulation models. In: Kuhl ME, M SN, Armstrong FB, Joines JA (eds) 2005 winter simulation conference, Orlando, Florida, USA, pp 130–143

155. International Towing Tank Conference (2011) ITTC Recommended Procedures and Guidelines 7.5-02-06-03: Validation of manoeuvring simulation models
156. Hwang W (2004) Guidelines for documenting math models for shiphandling simulator users not majored in hydrodynamics and math modeling. In: Workshop discussion at 31st IMSF annual general meeting, Antwerp, Belgium
157. Endo M (2005) Model documenting guidelines for ship-handling simulators. In: 35th annual general meeting (AGM) and international workshop conference of the international marine simulator forum (IMSF)
158. Crane CL (1979) Maneuvering trials of 278,000 DWT tanker in shallow and deep waters. SNAME Trans 87:251–283
159. Simonsen CD, Stern F (2003) Verification and validation of RANS maneuvering simulation of Esso Osaka: Effects of drift and rudder angle on forces and moments. Comput Fluids 32:1325–1356
160. Simonsen CD, Stern F (2005) RANS Maneuvering simulation of Esso Osaka with rudder and a body-force propeller. J Ship Res 49(2):98–120
161. International Towing Tank Conference (1999) ITTC Recommended procedures and guidelines 7.5-04-02-01: CFD, Resistance and flow benchmark database for CFD validation for resistance and propulsion
162. ITTC Manoeuvring Committee (1999) Final Report and Recommendations to the 22nd ITTC. In: 22nd International towing tank conference, Seoul, Korea and Shanghai, China, vol 1, pp 71–172
163. ITTC Manoeuvring Committee (2005) Final Report and Recommendations to the 24th ITTC. In: 24th International towing tank conference, Edinburgh, UK, vol 1, pp 137–197
164. Stern F, Agdrup K, Kim SY, Hochbaum AC, Rhee KP, Quadvlieg F, Perdon P, Hino T, Broglia R, Gorski J, (2009) Lessons learnt from the workshop on verification and validation of ship manoeuvring simulation methods - SIMMAN, (2008) In: International conference on marine simulation and ship maneuverability (MARSIM '09), Panama City, Panama
165. Stern F, Agdrup K, Kim SY, Hochbaum AC, Rhee KP, Quadvlieg F, Perdon P, Hino T, Broglia R, Gorski J (2011) Experience from SIMMAN 2008 - The first workshop on verification and validation of ship maneuvering simulation methods. J Ship Res 55(2):135–147
166. Quadvlieg F, Van Coevorden P (2003) Manoeuvring criteria: More than IMO A751 requirements alone! In: International conference on marine simulation and ship maneuverability (MARSIM '03), Kanazawa, Japan, vol 2, pp 1–8

Chapter 6
Maneuvering Modeling of Inland Vessels

The maneuverability of seagoing ships has been studied extensively, but several factors make it doubtful to directly apply the seagoing ship oriented research to inland vessels. These factors relate to the environmental conditions in which inland vessels operate and the particulars of the vessels. Given the impact factors on ship maneuverability and the differences between seagoing ships and inland vessels, Chap. 5 describes the extra impact factors to be addressed for inland vessel maneuverability and identifies the rudder configuration as a crucial difference. To analyze the impacts of the rudder configuration on ship maneuvering performance, the rudder induced forces need to be determined and addressed in this chapter.

6.1 Challenges for Mathematical Modeling of Inland Vessels

Thus far, research in the field of ship maneuverability has been primarily focused on seagoing ships [1]. However, interest in inland shipping is growing [2]. Generally, inland vessels have more complex arrangements than seagoing ships. An example of common inland vessel rudder and propulsion arrangements is shown in Fig. 6.1, which has a hull tunnel in front of ducted twin propellers and quadruple rudders. The complex arrangements also make studies on the maneuverability of inland vessels harder to tackle than those of seagoing ships. Furthermore, sailing in a complex navigation environment, inland vessels are expected to be more maneuverable than seagoing ships to ensure navigation safety. To ensure a good performance in practice, accurate methods of prediction are needed to link the design parameters to maneuvering performance. Additionally, relevant requirements on maneuverability should be established for designers and authorities.

To evaluate the maneuvering performance of ships, standard test maneuvers and related criteria are needed. maneuverability standards issued by the International Maritime Organization (IMO) [3, 4] are widely applied. These standards should be

© Springer Nature Switzerland AG 2020
J. Liu, *Mathematical Modeling of Inland Vessel Maneuverability Considering Rudder Hydrodynamics*, https://doi.org/10.1007/978-3-030-47475-1_6

Fig. 6.1 An inland vessel with a hull tunnel, ducted twin propellers, and quadruple rudders in the Rhine

applied to "ships of all rudder and propulsion types, of 100 m in length and over, and chemical tankers and gas carriers regardless of the length" [3, 4]. The type of ships (seagoing ships or inland vessels) is not clearly stated in the IMO standards. However, the applicable condition of the standards is described as deep, unconstrained water ($H/T > 4$). It is therefore rational to suppose that the IMO standards are intended for seagoing ships, but also valuable as guidance for inland vessels.

Regulations for inland vessels have been proposed by regional authorities [5–7] and classification societies [8], which define required evasive action and stopping capacity. Besides, China Classification Society [9] set up more test maneuvers, including turning circle, 15°/15° zigzag, stopping, pull-out, and course keeping tests. ITTC Manoeuvring Committee [10] gave a review of the standards in use for inland vessels, encouraging more test scenarios and criteria. According to Gray et al. [11], it is still doubtful if the standards lead to adequate maneuverability in shallow, restricted, and congested waterways. Therefore, in either case of inland vessels or seagoing ships, new test maneuvers and procedures for shallow water operations are demanded to predict and evaluate ship maneuverability [12, 13].

Considering the above-mentioned research challenges and gaps in knowledge, there are several open issues related to inland vessel maneuverability:

- **The adaptability of the previous research, intended for seagoing ships, to inland vessels has not been examined yet.** Existing research on ship maneuverability has been primarily performed for seagoing ships. However, inland vessels and seagoing ships are different in design options, rudder configurations, and navigation environment. These differences make it questionable to apply seagoing ship based results directly to inland vessels. It is worthwhile to study the critical differences between inland vessels and seagoing ships and their impacts on ship maneuvering performance.

- **Impacts of design choices in rudder configurations, namely the rudder profile, the rudder parameters, the number of rudders, and the spacing among multiple rudders, on ship maneuverability are not clear.** Current empirical formulas are established from the research on seagoing ships which commonly use a single NACA rudder while inland vessels may have more options. Even for seagoing ships, twin rudders and high-lift profiles are also applied owing to the ship enlargement and the requisite for improvement of ship maneuverability. Therefore, it is necessary to set up guidance for naval architects to make proper choices of rudder configurations.

- **Existing test maneuvers and criteria for inland vessels are not as elaborate as those for seagoing ships.** For seagoing ships, turning, zigzag, and stopping tests are mandatory while additional spiral (direct, reverse, and simplified), pull-out, and very small zigzag are suggested [3, 4]. For inland vessels, only tests of evasive action and stopping capacity are regulated [6]. Furthermore, the existing maneuvers are insufficient to present ship maneuvering performance in a complex navigation environment. More practical and handy maneuvers are needed to evaluate inland vessel maneuverability and compare the performance of various inland vessel designs.

Seagoing ships are commonly equipped with a single propeller and a single rudder while inland vessels more frequently feature multiple propellers and multiple rudders, particularly twin propellers and twin rudders. Furthermore, TPTR ships are preferred in shallow water owing to its superiority maneuvering performance [14]. Yoshimura and Sakurai [15] showed that a wide-width TPTR ship may have an improved turning ability in shallow water instead of a customarily worsened one for conventional SPSR ships. In general, TPTR vessels are appropriate for ships in shallow water, for instance, inland vessels, and large ships that require extra power and maneuverability, as an example, large seagoing container ships.

The interactions among multiple propellers and multiple rudders should be considered in maneuvering performance analysis. Khanfir et al. [14] noted that TPTR ships are normally wider than single-propeller single-rudder (SPSR) ships. These features, i.e. wide and blunt, are quite typical for inland vessels and may lead to significantly different maneuvering characteristics from seagoing ships. Previous studies have revealed the lack of experimental or theoretical research for multiple-propeller and multiple-rudder (MPMR) ships, including TPTR ships [15–18].

Besides studies on the mathematical modeling of MPMR ships [15, 17, 19–22], several attempts have been made to model the interaction between the hull and the rudder [14], the interaction between the propeller and the rudder [23], and the flow straightening effect of the hull and the propeller on the rudder [24]. Several studies have been published on the asymmetric maneuvering behavior of MPMR ships [25–27]. Modeling of the single-propeller twin-rudder system is a challenge as the inflow to the rudders is strongly dependent on the propeller action, the ship motion, and the twin-rudder interactions [28, 29]. Thus far, the rudder conflagrations have not been given great attention.

The applied methodology of study on inland vessel maneuverability in this monograph is based on the assumption that the maneuverability of a vessel can be judged by its maneuvering trajectories and criteria in specified test maneuvers. Owing to the availability of data, reference ships are taken from Chinese and Dutch fleets. The lengths of these ships are around 110 m. Only conventional self-propelled motor vessels are considered, while pusher-barge and towed-barge systems are not explicitly discussed. This monograph has the following research assumptions and boundaries:

- Shallow water effects on ship maneuverability and resistance are not explicitly studied.
- Service-speed maneuverability of the reference ships is evaluated while slow-speed maneuvering performance is not considered.
- Bow and stern thrusters are not considered.
- Hydrographic and meteorological conditions, including the current and the wind, are not considered.
- Channel characteristics, such as the width and the slope of the bank, are not included.
- Ship-ship interactions and ship-bank interactions are not included.
- Human factors are not included.

The above-mentioned assumptions limit the accuracy of the presented maneuvering models in the applications to slow-speed maneuvering in shallow water. However, the analyzed impact factors and constructed model structures hold. Further improvements can be achieved through an extensive study on hydrodynamic and efficiency coefficients of the ship hull and equipment (including propellers, rudders, bow and stern thrusters, and/or other types of active steering gears).

6.2 Reference Ships

As previously discussed in Chap. 2, ship particulars and rudder configurations affect ship maneuverability and thus influence the choices that are made in the development of the maneuvering model. This section introduces the applied reference ships. These ships are the single-propeller single-rudder (SPSR S1) seagoing KVLCC2 tanker, the single-propeller twin-rudder (SPTR R1) and twin-propeller quadruple-rudder (TPQR R2) standard inland vessels on the Rhine River, and the twin-propeller twin-rudder 6700 t bulk carrier (TPTR Y1) and 3500 t tanker (TPTR Y2) from the Yangtze River. Ship particulars and experimental results of the inland vessels are provided by Maritime Research Institute Netherlands (MARIN) [30] and Wuhan Rules and Research Institute, China Classification Society (CCS).

Table 6.1 presents the full-scale particulars of the reference ships. The main particular ratios of the two reference inland vessels from the Yangtze estuary (TPTR Y1

Table 6.1 Full-scale particulars of the reference ships

Ship model	SPSR S1	TPTR Y1	TPTR Y2	SPTR R1	TPQR R2
Kind of ship Environment	KVLCC2 tanker Sea	Inland bulk carrier Yangtze River	Inland tanker Yangtze River	Inland tanker Rhine River	Inland tanker Rhine River
Scale	45.71	24.27	22.82	18.00	18.00
V (ms^{-1})	7.97	5.00	5.00	4.47	4.47
L (m)	320.0	107.5	94.6	110.0	110.0
B (m)	58.0	19.2	17.2	11.4	11.4
T (m)	20.8	4.2	4.2	3.5	3.5
S_H (m)	27194	2784.30	2209.30	1866.11	1857.49
∇ (m)	312622	7561.70	5844.82	3876.09	3862.22
C_b (–)	0.8098	0.867	0.855	0.883	0.880
C_m (–)	0.9980	0.999	0.999	0.999	0.999
C_p (–)	0.8120	0.868	0.856	0.884	0.881
D_P (m)	9.86	2.50	2.35	1.80	1.80
P_P/D_P (–)	0.71	0.75	0.68	1.00	1.00
n_P (–)	1	2	2	1	2
n_R (–)	1	2	2	2	4
C_R (m)	0.71	3.05	2.60	1.80	1.80
A_R (m)	115.04	10.51	6.76	3.51	3.51
Λ_G (–)	1.83	1.13	1.00	1.08	1.08
y_{TR}/C_R (–)	–	3.8	4.0	0.89	0.92
y_{TU}/C_R (–)	–	–	–	–	1.75
Rudder profile	NACA 0018	NACA 0015	NACA 0015	Fishtail	Fishtail
Rudder endplate	No	Yes	No	Yes	Yes

and TPTR Y2) are different from those of the European inland vessels (SPTR R1 and TPQR R2) and common seagoing ships (SPSR S1). Furthermore, the water depth to the ship draft ratio (H/T) of the Yangtze River, especially in the downstream, is commonly much larger than that of the European waterways where H/T is typically around 1.4. Therefore, shallow-water effects are neglected for the reference Yangtze inland vessels while they might be important for European inland vessels.

Free-running test data of TPTR Y1 and TPTR Y2 were provided by Wuhan Rules and Research Institute, China Classification Society (CCS) to validate the maneuvering model of the reference inland vessels. These two reference ships are different in the hull form and the rudder configuration. TPTR Y1 fits a bulbous bow and a slender stern while TPTR Y2 features a normal bow and a blunt stern. In general, a bulbous bow reduces wave-making resistance over a narrow range of speed and draft, which results in higher speed for the same power or lower power for the same speed.

Fig. 6.2 The NACA 0015
profile with endplates of the
6700 t bulk carrier (TPTR
Y1)

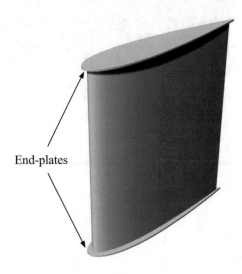

End-plates

    Most of the previous studies focused on the impacts of the bulbous bow on power-
ing while very few studies mentioned its impacts on ship maneuverability. Compared
to the particulars of TPTR Y1 (a bulbous bow, a slender stern, and rudders without
endplates), the features of TPTR Y2 (a normal bow, a blunt stern, and rudders with
endplates) are more typical for ships from the Yangtze River. Additionally, inland
vessels in Europe are more similar to the arrangements of TPTR Y2 than TPTR Y1.
    Regarding the rudder configuration, both vessels equip twin spade type NACA
0015 rudders. However, the rudders of TPTR Y1 have top and bottom endplates
while those of TPTR Y2 do not have endplates. The endplates are commonly applied
to inland vessels to improve the effective aspect ratios of the rudders and, further,
improve the ship maneuvering performance. Figure 6.2 shows the rudder profile and
the endplates.

6.3   Mathematical Models of Ship Motion

Mathematical models are used to investigate the ship maneuverability through computer simulations after which maneuvering performance can be evaluated with the
standards discussed in Chap. 7. In this section, these mathematical models and their
required inputs are discussed as a basis for the evaluation of those inputs in Sect. 5.5.
Initially, coordinate systems are defined to denote the ship trajectory and express the
forces and moments (Sect. 5.3). Newton's laws of motion can be applied to relate ship
movements to the acting forces and moments in the predefined domain (Sect. 5.4).
Afterward, Sect. 5.5 discussed expressions of forces and moments applied to solve
the ship motion equations.

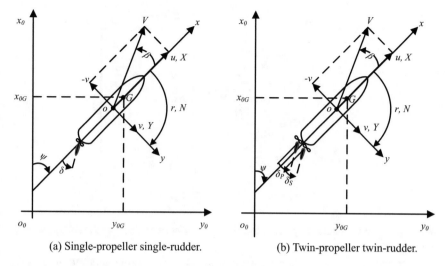

(a) Single-propeller single-rudder.   (b) Twin-propeller twin-rudder.

Fig. 6.3 Earth-fixed and body-fixed coordinate systems

To formulate the mathematical description of ship motions, coordinate systems should be established first. Two right-handed frames of axes are applied in bidirectional earth-fixed and body-fixed coordinate systems. The origin of the earth-fixed coordinate system is commonly located at the start point of the maneuvering simulation while the origin of the body-fixed coordinate system is set on midship or at the center of gravity. As highlighted by Yasukawa and Yoshimura [31], midship is more convenient than the center of gravity considering shipload conditions and thus applied as the origin of the body-fixed coordinate system in this monograph.

Figure 6.3 illustrates the static earth-fixed (o_0-$x_0y_0z_0$) and the dynamic body-fixed (o-xyz) coordinate systems for SPSR and TPTR ships as examples. The origin of o-xyz locates on midship. x, y, and z axes are positive to the bow of the ship, the starboard of the ship, and downwards of the water surface respectively. The ship position is determined by the center of gravity (G) of the ship in o_0-$x_0y_0z_0$, which is (x_G, 0, 0) in o-xyz for the applied three-dimensional model.

If not specified, parameters are defined on midship. Subscript G denotes variables at the center of gravity. Assuming the ship presented in Fig. 6.3 is maneuvering at forward speed (u) and lateral speed (v), the advance speed is $V = \sqrt{u^2 + v^2}$. The heading angle (ψ), longitudinal speed ($u_G = u$), and horizontal speed ($v_G = v + x_G r$) express the ship motion state in $o_0 - x_0y_0z_0$. The ship is turning with a rudder angle (δ) at yaw rate ($r = \dot{\psi}$) with a drift angle ($\beta = \arctan(-v/u)$).

In this monograph, prime symbols stand for non-dimensional parameters. Linear velocity (u, v), angular velocity (r), force (X, Y), moment (N), mass (m, m_x, m_y), and moment of inertia (I_z, J_z) are non-dimensionalized as follows:

$$\left.\begin{array}{r}
u', v' = \dfrac{u}{V}, \dfrac{v}{V} \\[2mm]
r' = \dfrac{rL}{V} \\[2mm]
X', Y' = \dfrac{X}{0.5\rho LTV^2}, \dfrac{Y}{0.5\rho LTV^2} \\[2mm]
N' = \dfrac{N}{0.5\rho L^2 TV^2} \\[2mm]
m', m'_x, m'_y = \dfrac{m}{0.5\rho L^2 T}, \dfrac{m_x}{0.5\rho L^2 T}, \dfrac{m_y}{0.5\rho L^2 T} \\[2mm]
I'_z, J'_z = \dfrac{I_z}{0.5\rho L^4 T}, \dfrac{J_z}{0.5\rho L^4 T},
\end{array}\right\} \qquad (6.1)$$

where ρ is the water density, L, is the ship length between perpendiculars, and T is the ship draft.

Dynamic equations for ships with different numbers of propellers and rudders are not that different. Yasukawa and Yoshimura [31] presented a standard MMG model for the KVLCC2 ship. Most of the previous studies on TPTR ship maneuverability used similar modular type MMG models [14–16, 32–34], which proves the usability of the MMG model for maneuvering prediction of TPTR ships in both shallow and deep water through adding dynamic characteristics [15]. Owing to different simplifications of the added mass and the location of the body-fixed coordinate system (COG or midship), the equations of motion are not entirely the same in the above-mentioned literature.

According to Yasukawa and Yoshimura [31], this monograph uses a general format of dynamic equations for all the reference ships as follows:

$$\left.\begin{array}{r}
(m + m_x)\dot{u} - (m + m_y)vr - x_G mr^2 = X_H + \displaystyle\sum_{i=1}^{n_P} X_P^i + \sum_{j=1}^{n_R} X_R^j \\[4mm]
(m + m_y)\dot{v} + (m + m_x)ur + x_G m\dot{r} = Y_H + \displaystyle\sum_{i=1}^{n_P} Y_P^i + \sum_{j=1}^{n_R} Y_R^j \\[4mm]
(I_z + x_G^2 m + J_z)\dot{r} + x_G m(\dot{v} + ur) = N_H + \displaystyle\sum_{i=1}^{n_P} N_P^i + \sum_{j=1}^{n_R} N_R^j,
\end{array}\right\} \qquad (6.2)$$

where the subscripts H, P, R indicate the hull, the propeller, and the rudder, the superscripts i and j refers to each propeller and each rudder respectively, n_P and n_R are the numbers of propellers and rudders respectively, m, m_x, m_y are the ship mass, the added mass in x-direction, and the added mass in y-direction, I_z and J_z are the moment of inertia and the added moment of inertia around the z-axis, u and v are ship forward and lateral speeds, r is ship yaw rate around midship, and the dot notation of u, v, and r denotes the derivative of each parameter. In Eq. 6.2, the interactions among the hull, the propeller, and the rudder are not explicitly expressed but included in the calculation of the forces induced by the propeller and the rudder.

6.4 (Added) Mass and (Added) Moment of Inertia

Ship mass (m) and moment of inertia (I_z) depend on the ship main particulars. Routinely, m is calculated as the following:

$$m = \rho \nabla, \tag{6.3}$$

where ∇ is the ship displacement volume. An estimation of the radius of gyration (i_z) is needed for I_z. Motora [35] estimated I_z as the follows:

$$\left. \begin{array}{l} i_z = 0.2536L \\ I_z = mi_z^2, \end{array} \right\} \tag{6.4}$$

which is widely used in current maneuvering studies. Since i_z is not very sensitive to the ship type, the Motora [35] method is taken for all the following maneuvering simulations.

According to Hooft and Nienhuis [36], the added mass (m_x and m_y) and moment (J_z) can be accurately estimated with the charts given by Motora [35, 37, 38]. Based on these charts, Zhou et al. [39] made regression formulas as follows:

$$\left. \begin{array}{l} \dfrac{m_x}{m} = \dfrac{1}{100}\left[0.398 + 11.97C_b\left(1 + 3.73\dfrac{T}{B}\right) + 2.89C_b\dfrac{L}{B}\left(1 + 1.13\dfrac{T}{B}\right) \right. \\ \left. \qquad + 0.175C_b\left(\dfrac{L}{B}\right)^2\left(1 + 0.54\dfrac{T}{B}\right) - 1.107\dfrac{L}{B}\dfrac{T}{B}\right] \\[2mm] \dfrac{m_y}{m} = 0.882 - 0.54C_b\left(1 - 1.6\dfrac{T}{B}\right) - 0.156(1 - 0.673C_b)\dfrac{L}{B} \\[2mm] \qquad + 0.826\dfrac{T}{B}\dfrac{L}{B}\left(1 - 0.678\dfrac{T}{B}\right) - 0.638C_b\dfrac{T}{B}\dfrac{L}{B}\left(1 - 0.669\dfrac{T}{B}\right) \\[2mm] j_z = \dfrac{L}{100}\left[33 - 76.85C_b(1 - 0.784C_b) + 3.43\dfrac{L}{B}(1 - 0.63C_b)\right] \\[2mm] J_z = mj_z^2. \end{array} \right\} \tag{6.5}$$

Originally for seagoing ships, Clarke et al. [40] proposed regression formulas based on Planar Motion Mechanism (PMM) tests for m_y' and J_z' as follows:

$$\left.\begin{aligned} m'_y &= \frac{\pi T}{L}\left[1 + 0.16C_b\frac{B}{T} - 0.51\left(\frac{B}{L}\right)^2\right] \\ J'_z &= \frac{\pi T}{L}\left[\frac{1}{12} + 0.017C_b\frac{B}{T} - 0.51\left(\frac{B}{L}\right)^2\right]. \end{aligned}\right\} \qquad (6.6)$$

Compared to m, m_x is relatively small and Clarke et al. [40] approximated m_x as 3–6% of m. The Zhou et al. [39] method and the Clarke et al. [40] method give comparable results. In this monograph, the Zhou et al. [39] method is chosen as it is based on widely used Motora [35, 37, 38] charts.

6.5 Propeller Forces and Moments

Section 6.5 presents the calculation procedures of the forces and moments induced by each propeller (X^i_P, Y^i_P, and N^i_P), which are expressed as the following:

$$\left.\begin{aligned} X^i_P &= (1 - t^i_P)T^i_P \\ Y^i_P &= 0 \\ N^i_P &= -y^i_P(1 - t^i_P)T^i_P, \end{aligned}\right\} \qquad (6.7)$$

where t_P is the propeller thrust deduction in maneuvering motions, T_P is the propeller thrust, and y_P is the relative position of the propeller to the centerline of the ship. According to Toxopeus [41, p. 30], the transverse force caused by a propeller rotating in an asymmetrical wake is relatively small and difficult to describe by the existing empirical formulas. Therefore, the propeller induced side force (Y^i_P) and its contribution to the propeller yaw moment (N^i_P) are assumed to be zero.

Following Eq. 6.7, the propeller thrust (T_P) is expressed as follows:

$$T^i_P = \rho(n^i)^2(D^i_P)^4 K^i_T, \qquad (6.8)$$

where n is the propeller revolution rate, D_P is the propeller diameter, and K_T is the propeller thrust coefficient. As the reference ships with multiple propellers have identical propellers rotating at the same constant number of revolutions per second, for simplicity in the following text, $n = n^i$ and $D_P = D^i_P$.

Using a usual format, K_T is calculated by second order polynomials of the propeller advance ratio (J_P) as:

$$K^i_T = k^i_2(J^i_P)^2 + k^i_1 J^i_P + k^i_0, \qquad (6.9)$$

where k_2, k_1, and k_0 are propeller open water characteristics in the representation of K_T. Additionally, these open water characteristics can be assumed to be constant against the water depth [15]. Furthermore, J_P is expressed as:

$$J_P^i = \frac{u(1 - w_P^i)}{nD_P},$$ (6.10)

where w_P is the wake fraction at the propeller position in maneuvering. Thus far, to resolve Eq. 6.7, the question becomes how to get the wake fraction (Sect. 6.5.1) and the propeller thrust deduction factor (Sect. 6.5.2) during maneuvering motions.

6.5.1 Propeller Wake Fraction

Lee et al. [16] concluded that the propeller's effective wake $(1 - w_P)$ during maneuvering and the flow straightening coefficient of the rudder in port and starboard turning of the ship are the unique parameters of TPTR ships. w_P^P and w_P^S may be different during maneuvering motions, but it is assumed that $w_P = w_P^i$ here due to lack of information. The wake fraction in maneuvering motions (w_P) is commonly estimated based on the wake fraction in straight moving (w_{P_0}) or simply assumed to be a constant that is identical to w_{P_0}.

For inland vessels, Kulczyk [42] indicated that for a centerline single propeller with single rudder $w_{P_0} \approx 0.235$ while for twin propellers with a single rudder behind each propeller $w_{P_0} \approx 0.32$. Harvald [43] proposed regression formulas for the wake fraction and the thrust deduction factor. Kristensen and Lützen [44] indicated that the Harvald [43] method may overestimate these values. Accordingly, Kristensen and Lützen [44] derived corrections based on the Harvald [43] method for tankers and bulk carriers. In this monograph, the Kulczyk [42] method is chosen as it is determined based on a database of inland vessels.

The wake fraction during maneuvering motions (w_P) is complicated and related to the drift angle, the yaw rate, the shape of the stern, the direction of the propeller rotation, and the propeller working load [15]. Kang et al. [28] concluded that the rudder type (single-rudder or twin-rudder) has little influence on w_{P_0} and w_P for a single-propeller ship. w_P can be estimated based on w_{P_0} considering the geometrical inflow angle to the propeller in maneuvering (β_P). β_P is defined as the following:

$$\beta_P^i = \beta - \frac{x_P^i}{L}r',$$ (6.11)

where x_P is longitudinal position of the propeller. Since $x_P = x_P^i$ for the reference inland vessels, $\beta_P = \beta_P^i$.

Lee et al. [16] showed that $1 - w_P$ of an SPSR ship changes systematically with β_P and reaches a minimum around $\beta_P = 0$. However, $1 - w_P$ of a TPTR ship shows a greatly asymmetric trend with the change of β_P. Hirano [45] expressed the relationship of w_P and w_{P_0} as:

$$\frac{w_P}{w_{P_0}} = \exp(-4\beta_P^2).$$ (6.12)

Furthermore, Matsumoto and Sueteru [46] described the relationship of the effective wake fraction $(1 - w_P)/(1 - w_{P_0})$ as the following:

$$\frac{1 - w_P}{1 - w_{P_0}} = 1 + (1 - \cos^2 \beta_P)(1 - |\beta_P|). \tag{6.13}$$

The above-mentioned methods are built up originally for SPSR ships. For simplicity, Yoshimura and Sakurai [15] assumed that $1 - w_P = 1 - w_{P_0}$ for each propeller of a TPTR ship. Kang et al. [28] described the procedure to obtain t_{P_0}, t_P, w_{P_0}, and w_P for an SPTR ship through model tests. This procedure is also usable for TPTR ships but the model tests at initial stage may not be possible. This monograph takes Eq. 6.12 as it is recommended by China Classification Society [9].

6.5.2 Propeller Thrust Deduction Factor

A propeller may have different deduction factors in straight moving (t_{P_0}) and maneuvering motions (t_P). Commonly, t_P is not significantly different from t_{P_0} for SPSR ships and presumed to be a constant [31]. According to Kulczyk [42], for a centerline single propeller with a single rudder $t_{P_0} \approx 0.27$ while for a twin-propeller ship with a single rudder behind each propeller $t_{P_0} \approx 0.2$. Lee et al. [16] showed that t_P of a TPTR ship is similar to that of a SPSR ship. Kang et al. [28] reported that the variation from t_{P_0} to t_P for SPTR ships appeared to be significant in the tested cases. However, additional tests on other ship types are needed to confirm this difference [28].

In this monograph, the assumption of the Yoshimura and Sakurai [15] method is followed that $t_P = t_{P_0} = t_P^i$. Similar to the calculation of w_P, Kristensen and Lützen [44] indicated that the Harvald [43] method may overestimate t_P and suggested corrections for tankers and bulk carriers. According to Hollenbach [47], the Holtrop and Mennen [48, 49] method is more suitable for twin-screw ships than the Harvald [43] method. Additionally, the Holtrop and Mennen [48, 49] method gives a similar result to the Kulczyk [42] method for the reference ships. Again, the Kulczyk [42] method for t_P is taken as it is the only method that is formed based on inland vessels.

6.6 Rudder Forces and Moments

Di Mascio et al. [22] pointed out that the rudder forces and moments are very difficult to evaluate as they are strongly affected by the complex flow in the stern region. Commonly, the interactions between the rudders are not considered [15]. However, the hydrodynamic characteristics of each rudder in multiple-rudder configurations are different due to the rudder profiles and the interaction between the twin rudders. Gim [18] showed that the distance between the twin rudders plays an important role in generating the side force and concluded that the critical distance between

Fig. 6.4 Rudder
hydrodynamic force
convention

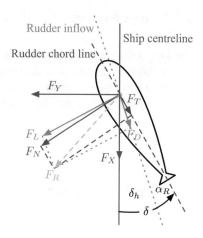

the rudders should be less than one chord length reducing the turbulence flow and vortices.

Kang et al. [28] concluded that the impact factors on twin-rudder performance are the inflow angle to each rudder, the interactions between the twin rudders, and the decrement of the inflow to the twin rudders. Furthermore, Kang et al. [28] observed that the effective wake fraction, the rudder inflow velocity, and flow straightening factor for the twin rudders are asymmetric during maneuvering. Kang et al. [25] investigated the inflow characteristics of each rudder on a single-propeller twin-rudder ship, showing that the inflow is not parallel to the ship centerline. This phenomenon leads to asymmetric maneuvering characteristics, reducing ship maneuverability.

Taking the origin on midship, total rudder forces and moments of a TPTR ship $(X_R^j, Y_R^j, \text{ and } N_R^j)$ are expressed as the following:

$$
\left.
\begin{aligned}
X_R^j &= -(1 - t_R^j)F_X^j \\
Y_R^j &= -(1 + a_H^j)F_Y^j \\
N_R^j &= -(x_R^j + a_H^j x_H^j)F_Y^j + y_R^j(1 - t_R^j)F_X^j,
\end{aligned}
\right\}
\tag{6.14}
$$

where F_X and F_Y are rudder force components in longitudinal and lateral directions, t_R is the steering resistance deduction factor, a_H is the rudder force increase factor, x_H is the longitudinal coordinate of the acting point of the additional lateral force, x_R and y_R are the longitudinal and lateral positions of the rudder. The applied force and angle conventions are illustrated in Fig. 6.4.

Section 6.6.1 introduces the method to calculate F_X and F_Y by lift and drag forces. Section 6.6.2 describes the applied rudder hydrodynamic coefficients based on the RANS results in Chap. 4. Section 6.6.3 computes the rudder inflow velocity components for the rudder force calculation. In the end, Sect. 6.6.4 discusses the applied values of the additional parameters, namely a_H, x_H, and t_R in Eq. 6.14.

6.6.1 Rudder Force Components

The longitudinal and lateral components of the rudder force (F_X and F_Y) can either be calculated by the rudder normal and tangential forces (F_N and F_T) or by the rudder lift and drag forces (F_L and F_D). In this monograph, the rudder forces and moments are calculated by using the rudder lift and drag forces as follows:

$$\left.\begin{aligned} F_X^j &= F_L^j \sin \delta_h^j + F_D^j \cos \delta_h^j \\ F_Y^j &= F_L^j \cos \delta_h^j - F_D^j \sin \delta_h^j, \end{aligned}\right\} \tag{6.15}$$

where δ_h^j is the hydrodynamic inflow angle of the rudder and expressed as:

$$\delta_h^j = \arctan\left(\frac{v_R^j}{u_R^j}\right), \tag{6.16}$$

where u_R^j and v_R^j are longitudinal and lateral components of the rudder inflow speed as $V_R^j = \sqrt{(u_R^j)^2 + (v_R^j)^2}$, which are discussed in Sect. 6.6.3.

The rudder hydrodynamic forces (F_L^j and F_D^j) are expressed as follows:

$$\left.\begin{aligned} F_L^j &= 0.5\rho A_R^j C_L^j (V_R^j)^2 \\ F_D^j &= 0.5\rho A_R^j C_D^j (V_R^j)^2, \end{aligned}\right\} \tag{6.17}$$

where C_N^j, C_T^j, C_L^j, and C_D^j are the rudder hydrodynamic coefficients. These coefficients are determined on the effective rudder angle (α_R), which is written as the following:

$$\alpha_R^j = \delta^j - \delta_h^j. \tag{6.18}$$

In this monograph, the general expressions for δ_h and α_R (Eqs. 6.16 and 6.18) are taken, which were applied by Yasukawa and Yoshimura [31], Yoshimura and Ma [50], Yoshimura and Masumoto [51] for SPSR ships, Kang et al. [28] for SPTR ships, and Yoshimura and Sakurai [15] for TPTR ships. Other expressions of α_R were described by Khanfir et al. [14], Nagarajan et al. [52] for different ship types. These methods may improve the accuracy of the prediction but require additional parameters that may not be available at the initial design stage.

6.6.2 Rudder Hydrodynamic Coefficients

Instated of using the general empirical formulas to estimate the rudder force coefficients, this monograph uses RANS methods to obtain the lift and drag coefficients

$(C_L$ and $C_D)$ of various rudder configurations in Chap. 4. A general format of C_L and C_D for the presented maneuvering model is written as follows:

$$\left.\begin{array}{l} C_L = k_P \left(\dfrac{\partial C_L^{2D}}{\partial \sin \alpha_R} \sin \alpha_R + C_{L_0}^{2D} \right) \dfrac{k_R \Lambda_G}{\Lambda_G + k_\Lambda} \\[4mm] C_D = k_P \left(\dfrac{\partial C_D^{2D}}{\partial \sin \alpha_R} \sin \alpha_R + C_{D_0}^{2D} \right) \dfrac{k_R \Lambda_G}{\Lambda_G + k_\Lambda}, \end{array}\right\} \tag{6.19}$$

where $\frac{\partial C_L^{2D}}{\partial \sin \alpha_R}$, $\frac{\partial C_D^{2D}}{\partial \sin \alpha_R}$, $C_{L_0}^{2D}$, and $C_{D_0}^{2D}$ are calculated with the new regression formulas that are proposed in Sect. 4.4, k_R is the impact factor of the rudder endplates, k_Λ is impact factor of the rudder aspect ratio, and k_P is the impact factor of the propeller slipstream,. These three impact factors are discussed in the following sections.

6.6.2.1 End Plate Effect

The rudders of the reference inland vessels (TPTR Y1, SPTR R1, and TPQR R2) are designed with top and bottom endplates. These endplates are configured to enlarge the effective aspect ratio of the rudder, therefore, improve the rudder effectiveness. To account the effect of the endplates on C_L and C_D, an additional amplify factor k_R is added in Eq. 6.19. However, no exact value of k_R was found in the literature. In this monograph, k_R is assumed to be 1.3 based on model test experience.

6.6.2.2 Effective Aspect Ratio Effect

The presented 2D simulations assume that the rudder has an infinite aspect ratio. As presented in Sect. 4.3, impacts in the aspect ratio increases the lift slope and reduces the induced drag. Thus, a larger span and aspect ratio is more efficient than a smaller aspect ratio. A large rudder aspect ratio might also achieve a good course keeping with a high lift curve and rapid response while a small rudder aspect ratio may enhance maneuvering performance with a large stall angle [53, p. 64]. Based on the lifting line theory, k_Λ in Eq. 6.19 should be 2. In this monograph, the empirical $k_\Lambda = 2.25$ is taken from the Fujii [54] method.

6.6.2.3 Propeller Slipstream Effect

Chapter 4 discussed the rudder hydrodynamics in 2D open water. In reality, ship rudders work in the propeller slipstream. Compared with the rudder working in the free stream, Molland and Turnock [55, 56] demonstrated a significant increase in the stall angle when the rudder is working behind a rotating propeller. The propeller slipstream may also have impacts on the rudder hydrodynamic coefficients depending on the propeller working load, the propeller revolution rate, and the arrangement of the propeller and the rudder. These impacts of the propeller-rudder interactions

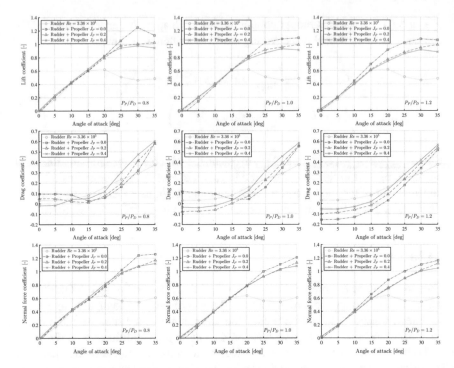

Fig. 6.5 Hydrodynamic coefficients of a rudder in open water or in propeller slipstream. Data are adapted from Nienhuis [59]

should be accounted for in the calculation of the rudder forces and moments for maneuvering simulations.

Rudder hydrodynamic characteristics tested in open water are different from those tested in the propeller slipstream. Oppenheim [57] found that the lift curve slope, the maximum lift coefficient, the stall angle, and the drag coefficient increase when the rudder is tested in the propeller slipstream. Kerwin et al. [58] compared the rudder hydrodynamics of a 20% flap rudder in propeller slipstream and uniform flow indicating that the rudder lift curve slope is increased by about 25% and this increase is constant over the tested range of longitudinal propeller locations. Nienhuis [59] performed tests for a rudder in open water or in propeller slipstream as compared in Fig. 6.5.

As shown in Fig. 6.5, the propeller slipstream extends the rudder stall angle from the range of 15° to 20° to the range of 30°–35°. The changes in J_P and P_P/D_P on the lift coefficients are comparatively small to the influence on the drag coefficients. Moreover, as the drag coefficients are relatively small values compared to the lift coefficients, the impacts of these changes are not very significant on the normal force coefficients. According to these results, it is assumed that the effect of the propeller slipstream primarily extends the stall angle without significantly affect the slope of the rudder hydrodynamic coefficients. Therefore, the impact factor of the propeller on the lift and drag slopes (k_P in Eq. 6.19) is assumed to be 1.0 in all the presented maneuvering simulations.

6.6.3 Rudder Inflow Velocity Components

Toxopeus [41, p. 30] concluded that the determination of the rudder inflow velocity and direction as a consequence of the drift angle (β), the yaw rate (r), and the propeller action is one of the most complicated aspects in determining the rudder forces and moments. The longitudinal and lateral inflow velocity components to each rudder (u_R and v_R) affect the amount of the lift and drag forces (Eq. 6.17) and the effective rudder angle (Eq. 6.18).

6.6.3.1 Longitudinal Component

According to Yoshimura and Sakurai [15], considering the relative position of the propeller (y_P) for multiple-propeller ships, u_R is expressed as:

$$u_R^j = (1 - w_R^j)(u - y_P^i r) \sqrt{\eta_R \left\{ 1 + \kappa_R \left[\sqrt{\left(1 + \frac{8K_T^i}{\pi (J^i)^2} \right)} - 1 \right] \right\}^2 + (1 - \eta_R)},$$

(6.20)

where κ_R is a constant that can be 0.5 [31] or 0.55 [51]. Kang et al. [28] assumed that κ_R is the same for both single-rudder and twin-rudder systems as they both have a single propeller. In this monograph, κ_R is estimated by the Yoshimura and Ma [50] method, which considers the ship particulars, as the following:

$$\kappa_R = 0.55 - 0.8 C_b \frac{B}{L}.$$

(6.21)

Additionally, w_R is the wake fraction at the position of the rudder, η_R is the ratio of the rudder area in the propeller slipstream (A_{RP}) to the area of the rudder movable area (A_R) and η_R is commonly estimated as:

$$\eta_R = \frac{A_{RP}}{A_R} \approx \frac{D_P}{B_R}.$$

(6.22)

Equation 6.22 is derived based on the assumption that the trailing edge of the rudder is always in the propeller slipstream. The effect of a part of the rudder that is out of propeller slipstream is not modeled and very little work is available in the literature. Quadvlieg [30] indicated that the rudders of inland vessels are always in the propeller slipstream, and this assumption is followed for the reference inland vessel.

The wake fraction at the position of the rudder during maneuvering motions (w_R) is commonly estimated based on the wake fraction of the rudder in straight moving (w_{R_0}), the wake fraction at the position of the propeller in straight moving (w_{P_0}) and maneuvering motions (w_P). According to Kijima et al. [60], w_R is expressed as the following:

$$w_R = w_{R_0} \frac{w_P}{w_{P_0}}. \tag{6.23}$$

Furukawa et al. [61] described w_R as the following:

$$w_R = \zeta_R w_{R_0} \exp\left[-4.0\left(\beta + \frac{r'}{2}\right)^2\right], \tag{6.24}$$

where ζ_R is an empirical coefficient and it is assumed to be 1.0 according to China Classification Society [9]. w_{R_0} can be estimated based on the wake fraction ratio (ε_R) as proposed by Kijima et al. [60],

$$\varepsilon_R = \frac{1 - w_{R_0}}{1 - w_{P_0}} = -156.2\left(C_b\frac{B}{L}\right)^2 + 41.6\left(C_b\frac{B}{L}\right) - 1.76. \tag{6.25}$$

In this monograph, the Kijima et al. [60] method (Eq. 6.25) is utilized to estimate ε_R for w_{R_0} and the Furukawa et al. [61] method (Eq. 6.24) for w_R. These two methods are pure empirical methods and have been widely applied on in previous studies for different ship types. Therefore, they are selected for the present work.

6.6.3.2 Lateral Component

According to Yasukawa and Yoshimura [31], Yoshimura and Masumoto [51], Yoshimura and Sakurai [15] for TPTR ships and SPSR ships, the lateral component of the rudder inflow velocity (v_R) is written as:

$$v_R^j = \gamma_R^j \beta_R^j, \tag{6.26}$$

where γ_R is the flow straightening factor due to the lateral speed of the ship (v). β_R is the effective inflow angle to the rudder in maneuvering and expressed as

$$\beta_R^j = \beta - \ell' r', \tag{6.27}$$

where $\ell' = \ell/L$ is the flow-straightening factor due to the yaw rate (r).

The flow straightening coefficient (γ_R) affects the inflow angle to the rudder the further influences the rudder induced forces and moments. Through wind-tunnel tests, Molland and Turnock [24] found that flow straightening effects depend on the type of upstream body, drift angles, and propeller thrust loading. Kang et al. [28] showed that γ_R depends on the effective rudder angle and the drift angle.

Kim et al. [33] indicated that the twin rudders have different inflow angles during maneuvering. The lee side rudder has larger flow straightening effects than the upwind side rudder. Meanwhile, the flow straightening coefficient γ_R for TPTR ship is smaller to that for the SPSR ship. For SPSR and TPTR ships, γ_R may be slightly asymmetric for port and starboard maneuvers [14, 16, 28] while for SPTR ships, these asymmetric phenomena become significant [14].

After reviewing the above-mentioned literature, it is assumed that γ_R for SPSR and MPMR ships are similar and not significantly different for port and starboard maneuvers. For SPSR ships, Kijima et al. [60] proposed such a formula:

$$\gamma_R = -22.2 \left(C_b \frac{B}{L} \right)^2 + 0.02 \left(C_b \frac{B}{L} \right) + 0.68. \tag{6.28}$$

For merchant ships and fishing vessels, Yoshimura and Masumoto [51] proposed:

$$\gamma_R = 2.06 C_b \frac{B}{L} + 0.14. \tag{6.29}$$

According to the experimental results from Yasukawa and Yoshimura [31], Yoshimura and Ma [50], Yoshimura and Masumoto [51] the possible range of γ_r is 0.4–0.6. Therefore, the Kijima et al. [60] method and the Ankudinov et al. [62] method may underestimate γ_R for the reference inland vessel. In this monograph, the Yoshimura and Masumoto [51] method is taken for γ_R as its prediction lies in the reasonable range of values.

Yasukawa and Yoshimura [31] described ℓ as an experimental constant from captive model tests. Kijima et al. [60] expressed $\beta_R = \beta - 2x'_R r'$ and thus $\ell' = 2x'_R \approx -1.0$ in general. Yoshimura and Ma [50] expressed ℓ' as the following:

$$\ell' = 1.7 C_b \frac{B}{L} - 1.2. \tag{6.30}$$

For fishing vessels and merchant ships, Yoshimura and Masumoto [51] indicated that $\ell' = -0.90$. Through trial and error, it is found that the maneuvering performance of the reference inland vessel is not very sensitive to ℓ'. The formula proposed by Yoshimura and Ma [50] is chosen in this monograph as it takes the main particulars of the ship into account.

6.6.4 Additional Parameters

In the following paragraphs, additional parameters in Eq. 6.14, namely the steering resistance deduction factor (t_R), the rudder force increase factor (a_H), and the longitudinal coordinate of the acting point of the additional lateral force (x_H), are discussed respectively.

6.6.4.1 Steering Resistance Deduction Factor

According to Kijima et al. [60], Matsumoto and Sueteru [46] for SPSR ships in deep water, the rudder steering resistance deduction (t_R) can be estimated as the following:

$$t_R = -0.28C_b + 0.45. \tag{6.31}$$

Koh and Yasukawa [63] indicated that the influence of B/T and y_P on t_R is not significant. Furthermore, Yoshimura [64] concluded that the variation of t_R with the change of water depth is negligible. Thus, Eq. 6.31 is taken for t_R in this monograph. Additionally, it is assumed that $t_R = t_R^j$.

6.6.4.2 Rudder Force Increase Factor

Based on the model tests carried out by Kose et al. [65], Yasukawa and Yoshimura [31] indicates the rudder force increase factor (a_H) has a common magnitude of 0.3–0.4, which means that the lateral force acting on the ship by steering increases about 30–40% larger than the rudder normal force component. According to Kijima et al. [60], a regression formula of a_H for SPSR seagoing ships in deep water is made as the following:

$$a_H = 2.0802C_b^2 - 0.6124C_b - 0.0569. \tag{6.32}$$

For inland vessels, Quadvlieg [30] proposed:

$$a_H = 0.627C_b - 0.153. \tag{6.33}$$

For fishing vessels and merchant ships, Yoshimura and Masumoto [51] calculated a_H as the following:

$$a_H = 3.6C_b\frac{B}{L}. \tag{6.34}$$

As shown by Khanfir et al. [14], a_H is dependent on B/T, C_b, and y_P while a_H may be changed significantly by the arrangement of the propellers and rudders. However, little information was found. From literature and experience [31, 52], a reasonable value of a_H should be around 0.4. In the presented simulations, the Quadvlieg [30] method is taken for a_H as it is the only method that is intended for inland vessels.

6.6.4.3 Longitudinal Coordinate of the Acting Point of the Additional Lateral Force

As stated by Khanfir et al. [17], the non-dimensional longitudinal coordinate of the acting point of the additional lateral force (x_H'), where $x_H' = x_H/L$, has a general value of –0.40. Furthermore, Khanfir et al. [14] indicated that x_H' has an almost constant value of -0.37, which is not significantly affected by B/T, C_b, and y_P. Therefore, it is reasonable to assume x_H is not much different for seagoing ships and inland vessels. As specified by Eloot [66], x_H' moves towards midships with decreasing under keel clearance leading to reduced turning ability in shallow water.

According to Kijima et al. [60], a regression formula for x_H' is made as the following:

$$x_H' = 9.5727C_b^2 - 8.0704C_b - 0.0618. \tag{6.35}$$

Lee and Shin [67] proposed formulas of x_H' for low-speed blunt ships with stern bulb and horn type rudders. After neglecting the terms that concern the stern bulb, the Lee and Shin [67] equation becomes:

$$x_H' = -6.054 + 58.18\frac{B}{L} - 148.44\left(\frac{B}{L}\right)^2. \tag{6.36}$$

In this monograph, the Lee and Shin [67] is selected to estimate x_H' as it concerns the ship main particulars.

6.7 Hull Forces and Moments

The hull induces longitudinal force (X_H), lateral force (Y_H), and yaw moment (N_H), which are commonly expressed in linear or non-linear functions of dimensional or non-dimensional dynamic parameters of the ship, such as u, $\dot{u}$, v, $\dot{v}$, r, $\dot{r}$, and β. The coefficients of such functions are the so-called hydrodynamic derivatives. Traditionally, these derivatives are derived from a series of model tests [14, 31]. Fast developments in CFD methods enable the approach of testing model-scale or full-scale ships in a numerical basin [34, 41, 68–71]. Experimental and numerical tests are expensive in money and time, therefore, commonly not available at the initial design stage.

The longitudinal component of the hull force (X_H) commonly refers to the resistance of the bare hull. X_H can be described as the component due to the straight moving $(X_H(u))$ and the component due to maneuvering $(X_H(v', r')$ or $X_H(\beta, r'))$. $X_H(u)$ can be obtained through model tests or estimated by the widely used Holtrop and Mennen method [72]. For simplicity or due to lack of data in the initial design stage, $X_H(v', r')$ or $X_H(\beta, r')$ can be neglected [9, 73]. In this monograph, the Holtrop and Mennen method [72] is used for $X_H(u)$.

Commonly, hydrodynamic coefficients obtained from model tests are used to express $X_H(v', r')$ or $X_H(\beta, r')$ [31, 60, 61, 74]. For initial design, empirical methods are easier and cheaper than experimental methods to apply for quick maneuverability prediction. Kang and Hasegawa [32], Yoshimura and Ma [50], Yoshimura and Masumoto [51] proposed regression formulas for the hydrodynamic coefficients in the expression of $X_H(\beta, r')$. Ankudinov and Jakobsen [75] provided regression

Table 6.2 Compared methods for the hull induced longitudinal force (X_H)

Case	Method for $X_H(u)$	Method for $X_H(\beta, r')$ or $X_H(v', r')$
No. 1	Holtrop [72]	Neglect $X_H(\beta, r')$ or $X_H(v', r')$
No. 2	Holtrop [72]	Yoshimura and Ma [50] for $X_H(\beta, r')$
No. 3	Holtrop [72]	Ankudinov and Jakobsen [75] for $X_H(v', r')$
No. 4	Holtrop [72]	Kang and Hasegawa [32] for $X_H(\beta, r')$
No. 5	Holtrop [72]	Yoshimura and Masumoto [51] for $X_H(\beta, r')$

Table 6.3 Compared methods for the hull induced lateral force and yaw moment (Y_H and N_H)

Case	Method for Y_H and N_H
No. 1	Kijima et al. [60]
No. 2	Matsunaga [74]
No. 3	Yoshimura and Ma [50]
No. 4	China Classification Society [9]
No. 5	Kijima and Nakiri [76]
No. 6	Kang and Hasegawa [32]
No. 7	Furukawa et al. [61]
No. 8	Yoshimura and Masumoto [51]

formulas to estimate $X_H(v', r')$. With or without $X_H(v', r')$ or $X_H(\beta, r')$, 5 methods for X_H are compared as listed in Table 6.2.

According to Yoshimura and Sakurai [15], the hydrodynamic derivatives of the hull, the propeller, and the rudder of a TPTR ship are not much different from those of an SPSR ship. Therefore, the existing regression formulas which were developed based on SPSR ships, including, but not limited to, China Classification Society [9], Furukawa et al. [61], Kang and Hasegawa [32], Kijima and Nakiri [76], Kijima et al. [60], Matsunaga [74], Yoshimura and Masumoto [51], are still applicable for TPTR ships. In this monograph, 8 methods are compared for Y_H and N_H as listed in Table 6.3.

To properly use the empirical methods for hull forces and moments, the applicable ranges of the ship main particulars, such as L/B, B/T, L/T, and C_b should be inspected as most of the empirical formulas are originally formulated based on databases of seagoing ships. Moreover, the dynamic parameters, such as the ship speed (V), the yaw rate (r'), and the drift angle (β) should also be examined.

Table 6.4 lists the static parameters of the reference inland vessels and the applicable ranges of the above-mentioned regression formulas. None of the listed 8 methods

Table 6.4 Parameter ranges of the reference ships and the existing regression formulas

Ship model	L/B	B/T	L/T	C_b
SPSR S1	5.99	2.57	15.38	0.81
TPTR Y1	5.59	4.57	25.64	0.87
TPTR Y2	5.50	4.10	22.53	0.87
SPTR R1	9.65	3.26	31.43	0.88
TPQR R2	9.65	3.26	31.43	0.88
Method	L/B	B/T	L/T	C_b
Kijima et al. [60]	4.51–6.89	2.38–4.09	13.66–25	0.52–0.84
Matsunaga [74]	4.51–6.89	2.38–4.09	13.66–25	0.52–0.84
Yoshimura and Ma [50]	2.60–5.20	2.17–2.70	5.65–14.05	0.57–0.66
China Classification Society [9]	5.71–16.67	3.5–6.0	20–100	–
Kijima and Nakiri [76]	2.60–5.20	2.17–2.70	5.65–14.05	0.57–0.66
Kang and Hasegawa [32]	5.00–6.13	2.43–3.31	13.70–18.18	0.78–0.83
Furukawa et al. [61]	4.51–6.89	2.38–4.09	13.66–25	0.52–0.84
Yoshimura and Masumoto [51]	2.60–7.10	2.17–4.00	5.65–28.40	0.51–0.65

perfectly covers the reference inland vessels. In general, the applicable range of the Kijima et al. [60] method, the Matsunaga [74] method, the Furukawa et al. [61], and the Kang and Hasegawa [32] method fit better than the other methods.

Section 6.7 has discussed the available regression methods for the hull forces and moments. Currently, all the regression formulas are made from databases of seagoing ships. No method has been specially generated for inland vessels. Furthermore, the parameter ranges of the reviewed methods do not fit the reference inland vessels as shown in Table 6.4. Tables 6.2 and 6.3 list the compared 5 methods for X_H and 8 methods for Y_H and N_H. 40 combinations of these methods are made and tested to find the most suitable regression methods for the reference inland vessels. Comparing simulation results of each method to free-running tests, a final selection is made in Sect. 6.7.

Through reviewing the existing methods for the maneuvering parameters, the preselected methods for the reference inland vessels are summarized in Table 6.5. These methods are first used to select the most suitable method for the hull forces and moments in Sect. 6.7 and then utilized for the maneuvering validation of the reference inland vessels in Sect. 6.8. Tables 6.2 and 6.3 list the tested methods for the longitudinal and lateral hull forces and moments. In total, 40 combinations of the

Table 6.5 Applied methods for the reference inland vessels

Parameter	Method	Equation
I'_z	Motora [35]	Equation 6.4
m'_x, m'_y, and J'_z	Zhou et al. [39]	Equation 6.5
w_{P_0}	Kulczyk [42]	Constant
w_P	Hirano [45]	Equation 6.12
t_P	Kulczyk [42]	Constant
u_R	Yoshimura and Sakurai [15]	Equation 6.20
κ_R	Yoshimura and Ma [50]	Equation 6.21
w_{R_0}	Kijima et al. [60]	Equation 6.25
w_R	Furukawa et al. [61]	Equation 6.24
v_R	Yasukawa and Yoshimura [31]	Equation 6.26
γ_R	Yoshimura and Masumoto [51]	Equation 6.29
ℓ'	Yoshimura and Ma [50]	Equation 6.30
t_R	Kijima et al. [60]	Equation 6.31
a_H	Quadvlieg [30]	Equation 6.33
x'_H	Lee and Shin [67]	Equation 6.36
C_L and C_D	Section 6.6.2	Equation 6.19

methods for X_H, Y_H, and N_H are compared against free-running tests based on the reference 6700 t bulk carrier (TPTR Y1) and the 6700 t tanker (TPTR Y2).

To compare the performance of each combination, the average absolute deviation of the simulated turning criteria (σ_T) is calculated as the following:

$$\sigma_T = 100\% \times \sum_{i=1}^{n_T} \left(\left| \frac{A_D^{Sim} - A_D^{Exp}}{A_D^{Exp}} \right| + \left| \frac{T_D^{Sim} - T_D^{Exp}}{T_D^{Exp}} \right| + \left| \frac{V_C^{Sim} - V_C^{Exp}}{V_C^{Exp}} \right| + \left| \frac{r_C^{Sim} - r_C^{Exp}}{r_C^{Exp}} \right| \right) \Big/ 24,$$

(6.37)

where A_D is the advance, T_D is the transfer, V_C is the speed in steady turn, r_C is the yaw rate in steady turn, and n_T is the number of the performed turning maneuvers, The superscripts 'Sim' and 'Exp' stand for 'Simulation' and 'Experiment'. Furthermore, the average absolute deviation of the simulated zigzag criteria (σ_Z) is written as:

$$\sigma_Z = 100\% \times \sum_{i=1}^{n_Z} \left(\left| \frac{\psi_{O1}^{Sim} - \psi_{O1}^{Exp}}{\psi_{O1}^{Exp}} \right| + \left| \frac{\psi_{O2}^{Sim} - \psi_{O2}^{Exp}}{\psi_{O2}^{Exp}} \right| + \left| \frac{t_{O1}^{Sim} - t_{O1}^{Exp}}{t_{O1}^{Exp}} \right| + \left| \frac{t_{O2}^{Sim} - t_{O2}^{Exp}}{t_{O2}^{Exp}} \right| \right) \Big/ 24,$$

(6.38)

where ψ_{O1} and ψ_{O2} are the first and second overshoot angles, t_{O1} and t_{O2} are time at first and second overshoot angles, and n_Z is the number of the performed zigzag maneuvers.

The benchmark free-running tests include 6 turning (15°, 25°, and 35° for starboard and port sides) and 6 zigzag (10°, 15°, and 20° for starboard and port sides) tests. The model-scale initial advance speed (u_0) is slightly different for each test maneuver as listed in Tables 6.6 and 6.7 for the two reference inland vessels. Addi-

Table 6.6 Model-scale initial speed of each test maneuver for the 6700 t bulk carrier (TPTR Y1)

Turning	15°	25°	35°	−15°	−25°	−35°
u_0 (ms^{-1})	1.09	1.08	1.07	1.14	1.20	1.07
Zigzag	10°/10°	15°/15°	20°/20°	−10°/−10°	−15°/−15°	−20°/−20°
u_0 (ms^{-1})	1.00	1.03	1.03	1.07	1.00	1.01

Table 6.7 Model-scale initial speed of each test maneuver for the 3500 t tanker (TPTR Y2)

Turning	15°	25°	35°	−15°	−25°	−35°
u_0 (ms^{-1})	1.08	1.07	1.06	1.16	1.15	1.05
Zigzag	10°/10°	15°/15°	20°/20°	−10°/−10°	−15°/−15°	−20°/−20°
u_0 (ms^{-1})	1.04	1.06	1.03	1.06	1.04	0.59

tionally, the propeller revolution rates (n) are constants of 20.2 and 17.7 s^{-1} for the 6700 t bulk carrier and the 3500 t tanker respectively.

Using 40 different combinations of methods for X_H, Y_H, and N_H, the average absolute deviation of the simulated criteria in turning and zigzag maneuvers (σ_T with Eq. 6.37 and σ_Z with Eq. 6.38) are calculated and compared in Table 6.8. X_H No. in the first row refers to the methods listed in Table 6.2 while Y_H and N_H No. in the first column refers to the methods listed in Table 6.3. Some combinations in Table 6.8 fail to resolve the maneuvering equations and a few methods do not give reasonable results, which are indicated as 'Fail', due to inaccurate estimation of the hull forces and moments.

As the tested overshoot angles are actually small values, σ_Z is larger than σ_T in Table 6.8. It is supposed that the model with the smallest σ_T and comparable σ_Z is the most suitable combination. Therefore, the combination of X_H No. 5 with Y_H and N_H No. 1 gives the best fit in the prediction of turning and zigzag maneuvers for the 6700 t bulk carrier. To verify the selection, a similar procedure is carried out for the 3500 t tanker as shown in Table 6.9. Again, the combination of X_H No. 5 with Y_H and N_H No. 1 performs best for the 3500 t tanker. Thus, the selected combination is validated.

Using X_H No. 1 for X_H and different methods for Y_H and N_H, the steady yaw rates of the reference inland vessels in various turning maneuvers are simulated and compared with the experiment values as shown in Fig. 6.6. Y_H and N_H No. 1 can give the best fit among the tested methods, which proves the usability of this method to predict the steady maneuvering characteristics of the two reference inland vessels.

Through the comparison, it is determined to use the Holtrop and Mennen [48, 49] method for $X_H(u)$ and the Yoshimura and Masumoto [51] for $X_H(\beta, r')$ (X_H No. 5 in Table 6.2), therefore, $X_H = X_H(u) + X_H(\beta, r')$. Furthermore, Y_H and N_H is calculated by the Kijima et al. [60] method (Y_H and N_H No. 1 in Table 6.3). This combination is used to perform the maneuvering simulations in the following sections as it gives the best fit.

Table 6.8 Average absolute deviation of the 6700 t bulk carrier using different combinations of methods for hull forces and moments in turning and zigzag maneuvers

	X_H No. 1		X_H No. 2		X_H No. 3		X_H No. 4		X_H No. 5	
	σ_T (%)	σ_Z (%)	σ_T (%)	σ_Z (%)	σ_T (%)	σ_Z (%)	σ_T (%)	σ_Z (%)	σ_T (%)	σ_Z (%)
Y_H and N_H No. 1	6.49	10.59	7.03	10.62	Fail	11.15	12.86	10.57	**4.89**	**10.67**
Y_H and N_H No. 2	34.53	60.66	35.34	61.23	Fail	49.41	Fail	59.69	33.15	61.59
Y_H and N_H No. 3	104.74	38.18	32.81	38.48	Fail	33.40	Fail	37.58	35.44	38.84
Y_H and N_H No. 4	29.38	17.37	20.32	17.23	Fail	17.49	40.32	17.36	22.01	17.38
Y_H and N_H No. 5	21.27	46.39	20.23	46.96	Fail	36.25	24.87	45.04	19.98	47.40
Y_H and N_H No. 6	11.71	10.09	8.61	10.13	Fail	10.75	19.66	10.13	8.38	10.13
Y_H and N_H No. 7	21.46	57.77	19.81	58.55	Fail	42.53	25.22	55.36	19.37	59.13
Y_H and N_H No. 8	8.14	13.46	7.56	13.52	Fail	12.53	14.96	13.22	4.86	13.68

6.8   Validation of the Maneuvering Model

Section 6.8 introduces the validation of the presented maneuvering model. Section 6.8.1 validates the mathematical model with a seagoing KVLCC2 tanker (SPSR S1) as most of the applied parameters are from experimental tests. Sections 6.8.2 and 6.8.3 validate the integrated maneuvering model with two reference inland vessels (TPTR Y1 and TPTR Y2), which are carried out with the selected empirical methods (from Sects. 6.4–6.7) and the RANS results in Chap. 4.

Table 6.9 Average absolute deviation of the 3500 t tanker using different combinations of methods for hull forces and moments in turning and zigzag maneuvers

	X_H No. 1		X_H No. 2		X_H No. 3		X_H No. 4		X_H No. 5	
	σ_T (%)	σ_Z (%)	σ_T (%)	σ_Z (%)	σ_T (%)	σ_Z (%)	σ_T (%)	σ_Z (%)	σ_T (%)	σ_Z (%)
Y_H and N_H No. 1	9.69	29.81	8.45	29.96	36.76	27.89	14.10	29.51	**8.14**	**30.07**
Y_H and N_H No. 2	31.69	97.79	31.91	98.51	Fail	82.75	33.19	95.54	31.58	99.09
Y_H and N_H No. 3	51.23	132.15	42.07	133.87	Fail	103.16	Fail	124.65	43.85	135.66
Y_H and N_H No. 4	34.99	33.01	28.10	32.97	Fail	34.19	46.18	33.18	29.15	33.05
Y_H and N_H No. 5	30.85	162.12	30.66	165.43	Fail	117.01	33.16	152.25	30.55	167.73
Y_H and N_H No. 6	15.14	46.74	13.37	47.01	Fail	40.32	19.95	45.45	13.59	47.30
Y_H and N_H No. 7	32.48	248.50	32.08	259.67	Fail	153.89	34.64	222.69	32.06	262.73
Y_H and N_H No. 8	10.79	24.09	8.16	24.11	40.65	23.56	17.13	24.25	8.35	24.22

6.8.1 Validation Using the Seagoing KVLCC2 Tanker

The seagoing KVLCC2 (SPSR S1) tanker is utilized as the reference ship to validate the mathematical model (Table 6.1). The KVLCC2 tanker has a single spade-type rudder with a NACA 0018 profile. The free-running model tests carried out by the Maritime Research Institute Netherlands (MARIN) are used as reference [77, 78]. The scale factor is 45.7. Simulations are performed with the model-scale ship particulars as presented in Table 6.10. Then, both results of the model-scale simulations and experiments are transformed to full-scale and compared.

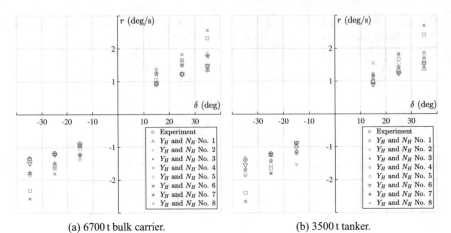

(a) 6700 t bulk carrier.   (b) 3500 t tanker.

Fig. 6.6 Comparison of the simulated and tested yaw rate (r) of the 6700 t bulk carrier and the 6700 t bulk carrier in various turning maneuvers

Table 6.10 Applied parameters for the KVLCC2 tanker. *Source* Yasukawa and Yoshimura [31]

L	7.0	B_{wl}	1.1688	T	0.455
∇	3.2724	C_b	0.8098	x_G	0.244
m'_x	0.022	m'_y	0.223	J'_z	0.011
u_0	1.179				
R'_0	0.022	Y'_v	−0.315	N'_v	−0.137
X'_{vv}	−0.040	Y'_r	0.083	N'_r	−0.049
X'_{vr}	0.002	Y'_{vvv}	−1.607	N'_{vvv}	−0.030
X'_{rr}	0.011	Y'_{vvr}	0.379	N'_{vvr}	−0.294
X'_{vvvv}	0.771	Y'_{vrr}	−0.391	N'_{vrr}	0.0550
		Y'_{rrr}	0.008	N'_{rrr}	−0.013
D_P	0.216	x'_P	−0.48	n	10.4
k_0	0.2931	k_1	−0.2753	k_2	−0.1385
t_P	0.220	w_{P_0}	0.40		
A_R	0.0539	Λ_G	1.827	x'_R	−0.50
t_R	0.387	a_H	0.312	x'_H	−0.464
C_1	2.0	$C_2\ (\beta_P > 0)$	1.6	$C_2\ (\beta_P < 0)$	1.1
$\gamma_R\ (\beta_R < 0)$	0.395	$\gamma_R\ (\beta_R > 0)$	0.640	ℓ'_R	−0.710
ε_R	1.09	κ_R	0.50	$\dot{\delta}$	15.8

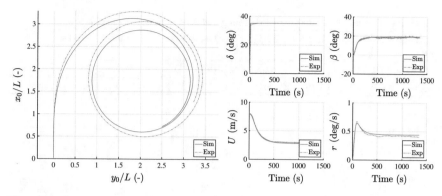

Fig. 6.7 Validation of the KVLCC2 tanker in the 35° turning maneuver

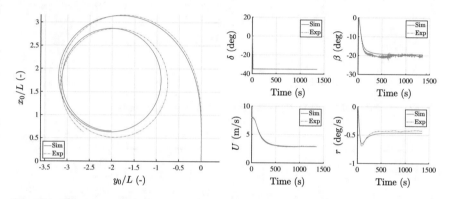

Fig. 6.8 Validation of the KVLCC2 tanker in the −35° turning maneuver

6.8.1.1 Validation of the Seagoing KVLCC2 Tanker in Turning Maneuvers

Figures 6.7 and 6.8 compare the simulated trajectories and histories to the free-running tests of 35° and −35° turning maneuvers [77, 78]. The simulated turning trajectories agree well with the free-running tests. The starboard side turning simulation fits better than the port side simulation. It shows that the maneuvering model underestimates the asymmetry motions of the port and starboard sides maneuvers. Table 6.11 compares the turning indices of the simulated results (Sim) with the free-running tests (Exp) [77, 78].

6.8.1.2 Validation of the Seagoing KVLCC2 Tanker in Zigzag Maneuvers

Figures 6.9, 6.10, 6.11 and 6.12 show the trajectories and histories in 10°/10°, −10°/−10°, 20°/20°, and −20°/−20° zigzag maneuvers respectively. The simulated results

Table 6.11 Comparison of the simulated turning and the tested turning indices

Turning criteria	Sim (–)	Exp (–)	Δ (%)	Turning criteria	Sim (–)	Exp (–)	Δ (%)
A'_D ($\delta = 35°$)	3.07	3.07	0.00	A'_D ($\delta = -35°$)	3.09	2.98	3.69
T'_D ($\delta = 35°$)	3.15	3.33	−5.41	T'_D ($\delta = -35°$)	3.14	3.09	1.62

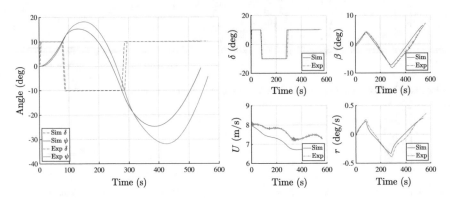

Fig. 6.9 Validation of the KVLCC2 tanker in the 10°/10° zigzag maneuver

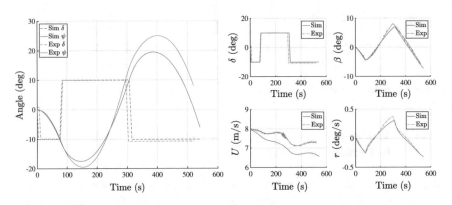

Fig. 6.10 Validation of the KVLCC2 tanker in the –10°/–10° zigzag maneuver

roughly agree with the experimental ship motions in zigzag maneuvers. A larger speed drop is observed in the simulated time histories than the experiments. The histories of the rudder angle (δ), the drift angle (β), and the turning rate (r) agree well. Table 6.12 compares the simulated overshoot angles to the benchmarks. The maneuvering indices are underestimated. The largest differences are about 30% in the first and second overshoot angles. It is actually very difficult to predict accurate overshoot angles in degrees. Here, the underestimation is mainly induced by the larger speed decrease, which may be caused by a larger hull damping force calculated in the maneuvering model.

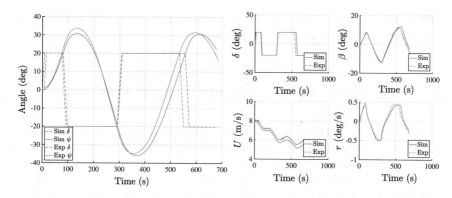

Fig. 6.11 Validation of the KVLCC2 tanker in the 20°/20° zigzag maneuver

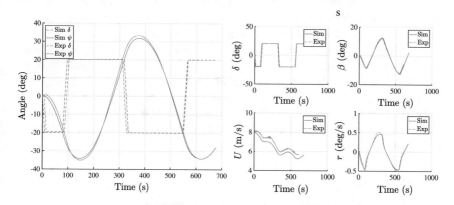

Fig. 6.12 Validation of the KVLCC2 tanker in the –20°/–20° zigzag maneuver

Table 6.12 Comparison of the simulated overshoot angles and the tested overshoot angles

Zigzag criteria	Sim (deg)	Exp (deg)	Δ (%)	Zigzag criteria	Sim (deg)	Exp (deg)	Δ (%)
ψ_{O1} (10°/10°)	5.3	7.9	–33.2	ψ_{O1} (–10°/–10°)	7.5	9.3	–19.1
ψ_{O2} (10°/10°)	14.7	21.6	–31.9	ψ_{O2} (–10°/–10°)	9.5	14.7	–35.4
ψ_{O1} (20°/20°)	10.8	13.3	–19.0	ψ_{O1} (–20°/–20°)	14.1	14.7	–3.9

Thus far, the applied mathematical model, which is a combination of the standard MMG model [31] and RANS-based rudder normal force coefficients, is validated with free-running tests. The model can capture the maneuvering motions and related parameters. Underestimations of the asymmetry motions are observed in turning maneuvers. Apart from the first overshoot angle of –20°/–20° zigzag test, the overshoot angles are underestimated due to a larger speed decrease simulated in the maneuvering model than experiments. Since the rudder forces and moments are calculated based on a specific profile, the model is applicable to investigate the impacts of rudder profiles on maneuvering performance.

6.8.2 Validation Using the Inland 6700 t Bulk Carrier

Maneuvering simulations of the 6700 t bulk carrier (TPTR Y1) are performed for $15°$, $25°$, and $35°$ turning maneuvers and $10°/10°$, $10°/10°$, and $10°/10°$ zigzag maneuvers on both both port and starboard sides. In the following sections, simulations are performed with model-scale parameters. Then, the simulated and tested results are scaled and compared in full-scale. The simulated and tested results of the trajectories, the rudder angle, the drift angle, the heading angle, the ship speed, and the yaw rates are compared. The drift angles of the experiment are not available. Here, only the simulated drift angles are shown to provide full information on the simulations. Table 6.13 presents the parameters used in the simulations.

As discussed in Sect. 4.2, the interaction between the rudders and shallow water also influences C_L and C_D. However, the spacing between the twin rudders of the reference inland vessels (TPTR Y1 and TPTR Y2) is rather large ($3.81 C_R$). It is assumed no significant interaction effect on C_L and C_D of each rudder. The shallow water effect on the rudder itself only applies to extremely small under keel clearance (Sect. 4.3), therefore, it is neglected. In the end, applied C_L and C_D for TPTR Y1 are described as follows:

Table 6.13 Applied parameters in the simulations of the 6700 t bulk carrier

L	4.429	B	0.791	T	0.173	∇	0.5290
S	4.665	C_b	0.867	x_G	0.009	m'_x	0.0178
m'_y	0.155	I'_z	0.021	J'_z	0.013	δ	25
$X'_{\beta\beta}$	-0.0019	$X'_{\beta r}$	-0.0608	X'_{rr}	0.0665	$X'_{\beta\beta\beta\beta}$	0.0665
Y'_β	0.3395	Y'_r	0.0973	$Y'_{\beta\beta}$	0.5727	Y'_{rr}	-0.0050
$Y'_{\beta rr}$	0.1731	$Y'_{\beta\beta r}$	-0.3656	N'_β	0.0781	N'_r	-0.0361
$N'_{\beta\beta}$	0.0381	N'_{rr}	-0.0126	$N'_{\beta rr}$	-0.0448	$N'_{\beta\beta r}$	-0.1295
D_P	0.110	w_{P_0}	0.32	t_P	0.2	x_P	-2.1220
y_P^S	-0.240	y_P^P	0.240	k_2^S	-0.1075	k_1^S	-0.3507
k_0^S	0.3329	k_2^P	-0.1295	k_1^P	-0.3269	k_0^P	0.3307
A_R	0.010	ε_R	0.9364	w_{R_0}	0.3633	C_R	0.126
B_R	0.142	Λ_G	1.129	x_R	-2.215	y_R^S	0.240
y_R^P	-0.240	t_R	0.2072	a_H	0.3906	x'_H	-0.3980
γ_R	0.4590	η_R	0.7752	ℓ'_R	-0.937	κ_R	0.426

$$C_L = 6.175 \sin \alpha_R \frac{1.3 \Lambda_G}{\Lambda_G + 2.25}$$
$$C_D = 0.032 \sin \alpha_R \frac{1.3 \Lambda_G}{\Lambda_G + 2.25}. \left.\right\}$$

(6.39)

6.8.2.1 Validation of the Inland 6700 t Bulk Carrier in Turning Maneuvers

Using the methods in Table 6.5 and the parameters in Table 6.13, the trajectories and time histories of simulated and tested motion parameters in the 15°, −15°, 25°, −25°, 35°, and −35° turning maneuvers are presented in Figs. 6.13, 6.14 , 6.15, 6.16, 6.17 and 6.18 respectively. Comparing the free-running results of various turning maneuvers, the starboard side A'_D and T'_D are larger than those of the port side.

Table 6.14 compares the non-dimensional advance (A'_D), the non-dimensional tactical diameter (T'_D), and the relative deviation in percentage (Δ) of the simulated

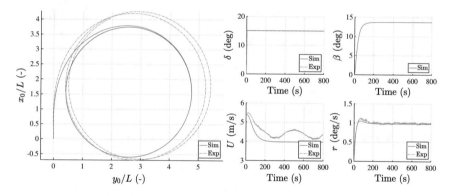

Fig. 6.13 Validation of the 6700 t bulk carrier in the 15° turning maneuver

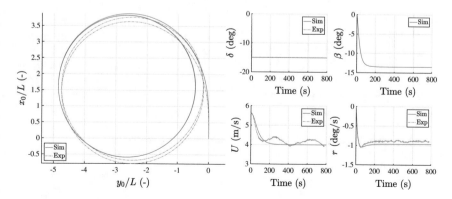

Fig. 6.14 Validation of the 6700 t bulk carrier in the −15° turning maneuver

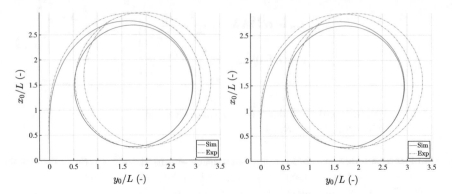

Fig. 6.15 Validation of the 6700 t bulk carrier in the 25° turning maneuver

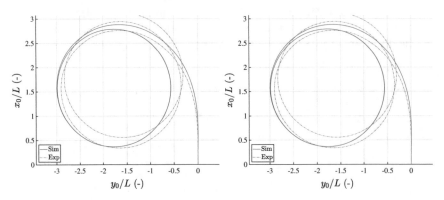

Fig. 6.16 Validation of the 6700 t bulk carrier in the −25° turning maneuver

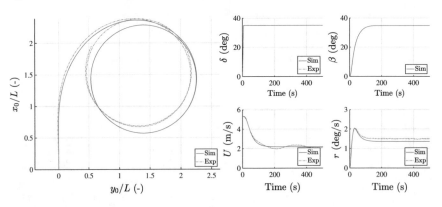

Fig. 6.17 Validation of the 6700 t bulk carrier in the 35° turning maneuver

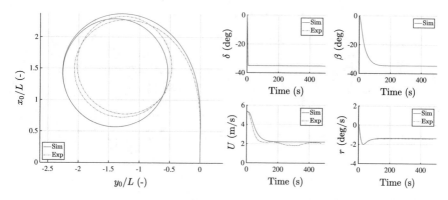

Fig. 6.18 Validation of the 6700 t bulk carrier in the −35° turning maneuver

Table 6.14 Comparison of the simulated and tested turning indices of the 6700 t bulk carrier

Turning criteria	Sim (–)	Exp (–)	Δ (%)	Turning criteria	Sim (–)	Exp (–)	Δ (%)
A'_D ($\delta =15°$)	3.74	4.08	−8.33	A'_D ($\delta =-15°$)	3.82	3.59	6.41
T'_D ($\delta =15°$)	4.71	5.08	−7.28	T'_D ($\delta =-15°$)	4.79	4.72	1.48
A'_D ($\delta = 25°$)	2.68	2.83	−5.30	A'_D ($\delta =-25°$)	2.79	2.68	4.10
T'_D ($\delta =25°$)	2.85	3.02	−5.63	T'_D ($\delta =-25°$)	2.89	2.84	1.76
A'_D ($\delta = 35°$)	2.23	2.27	−1.76	A'_D ($\delta =-35°$)	2.23	2.18	2.29
T'_D ($\delta = 35°$)	2.11	2.05	2.93	T'_D ($\delta =-35°$)	2.11	1.99	6.03

and tested turning maneuvers. Asymmetry behavior is found in the free-running trajectories while it is less obvious in the simulation results. Furthermore, the simulated A'_D and T'_D are mostly smaller on the port side while larger on the starboard side than the tested A'_D and T'_D.

6.8.2.2 Validation of the Inland 6700 t Bulk Carrier in Zigzag Maneuvers

The time histories of simulated and tested motion parameters in the 10°/10°, −10°/−10°, 15°/15°, −15°/−15°, 20°/20°, and −20°/−20° zigzag maneuvers are presented in Figs. 6.19, 6.20, 6.21, 6.22, 6.23 and 6.24 respectively. Table 6.15 compares the first overshoot angle (ψ_{O1}) and the second overshoot angle (ψ_{O2}) of the simulated and tested zigzag maneuvers. The simulated ψ_{O1} and ψ_{O2} are mostly larger than the tested ψ_{O1} and ψ_{O2}.

Comparing the results of port and starboard sides, the asymmetric characteristics of the simulated overshoot angles are less significant than the tested ones. The differences in the simulated and tested results are primarily caused by the divergence of the simulated and tested ship speed. The larger speed drop of the simulated results may be caused by the inaccurate prediction of the hull forces.

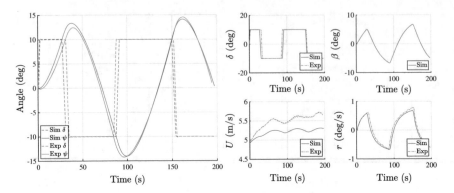

Fig. 6.19 Validation of the 6700 t bulk carrier in the 10°/10° zigzag maneuver

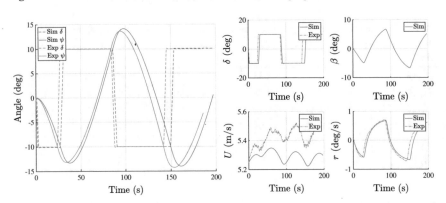

Fig. 6.20 Validation of the 6700 t bulk carrier in the −10°/−10° zigzag maneuver

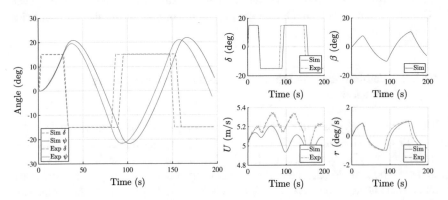

Fig. 6.21 Validation of the 6700 t bulk carrier in the 15°/15° zigzag maneuver

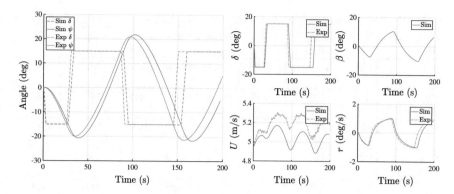

Fig. 6.22 Validation of the 6700 t bulk carrier in the −15°/−15° zigzag maneuver

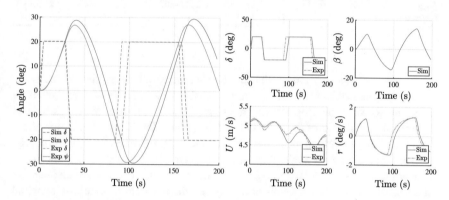

Fig. 6.23 Validation of the 6700 t bulk carrier in the 20°/20° zigzag maneuver

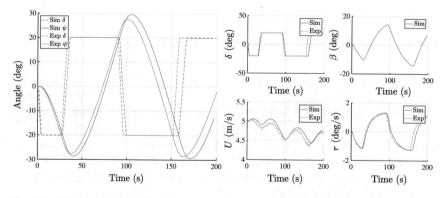

Fig. 6.24 Validation of the 6700 t bulk carrier in the −20°/−20° zigzag maneuver

Table 6.15 Comparison of the simulated and the tested overshoot angles of the 6700 t bulk carrier

Zigzag criteria	Sim (deg)	Exp (deg)	Δ (%)	Zigzag criteria	Sim (deg)	Exp (deg)	Δ (%)
ψ_{O1} $(10°/10°)$	3.36	2.43	38.27	ψ_{O1} $(-10°/-10°)$	3.30	3.16	4.43
ψ_{O2} $(10°/10°)$	4.02	4.17	-3.60	ψ_{O2} $(-10°/-10°)$	4.15	3.65	13.70
ψ_{O1} $(15°/15°)$	5.82	4.49	29.62	ψ_{O1} $(-15°/-15°)$	5.70	4.99	14.23
ψ_{O2} $(15°/15°)$	6.75	6.65	1.50	ψ_{O2} $(-15°/-15°)$	6.81	5.93	14.84
ψ_{O1} $(20°/20°)$	8.75	6.74	29.82	ψ_{O1} $(-20°/-20°)$	8.60	7.66	12.27
ψ_{O2} $(20°/20°)$	9.54	9.33	2.25	ψ_{O2} $(-20°/-20°)$	9.62	7.66	25.59

The simulation study shows that the prediction is quite sensitive to the wake fractions (w_P and w_R) and the flow straightening factor (γ_R). The wake fractions affect the inflow speed of the propeller and the rudder which, influence the magnitude of the forces and moments of the propeller and the rudder. The flow straightening effect affects the lateral component of the rudder inflow (v_R). Furthermore, it influences the hydrodynamic inflow angle of the rudder (δ_h) and the effective rudder angle (α_R) as shown in Eqs. 6.16, 6.18 and 6.26. In addition, the model-scale simulations are quite sensitive to the initial status of u_0 and n, which should be carefully matched with the experiments to perform the simulations.

6.8.3　Validation Using the Inland 3500 t Tanker

To further validate the proposed integrated model for inland vessels, this section presents maneuvering simulations of the 3500 t tanker (TPTR Y2). Table 6.16 presents the parameters used in the simulations. The Holtrop [72], Holtrop and Mennen [49] method is used for $X_H(u)$, the Yoshimura and Masumoto [51] method is used for $X_H(\beta, r')$, and the Kijima et al. [60] method is used for Y_H and N_H. Applied methods for maneuvering are listed in Table 6.5. Considering the rudder profile, the spacing between the twin rudders, and the endplates, the applied C_L and C_D for TPTR Y2 are written as follows,

$$\left.\begin{array}{l} C_L = 6.175 \sin \alpha_R \dfrac{\Lambda_G}{\Lambda_G + 2.25} \\[2mm] C_D = 0.032 \sin \alpha_R \dfrac{\Lambda_G}{\Lambda_G + 2.25}. \end{array}\right\} \tag{6.40}$$

Table 6.16 Applied parameters in the simulations of the 3500 t tanker

L	4.146	B	0.754	T	0.184	∇	0.492
S	4.244	C_b	0.855	x_G	0.011	m'_x	0.0194
m'_y	0.1708	I'_z	0.02	J'_z	0.0125	$\dot\delta$	25
$X'_{\beta\beta}$	−0.0012	$X'_{\beta r}$	−0.0462	X'_{rr}	0.0663	$X'_{\beta\beta\beta\beta}$	0.0613
Y'_β	0.3571	Y'_r	0.0973	$Y'_{\beta\beta}$	0.5885	Y'_{rr}	0.0016
$Y'_{\beta rr}$	0.2105	$Y'_{\beta\beta r}$	−0.3370	N'_β	0.0888	N'_r	−0.0401
$N'_{\beta\beta}$	0.0320	N'_{rr}	−0.0123	$N'_{\beta rr}$	−0.0543	$N'_{\beta\beta r}$	−0.1292
D_P	0.11	w_{P_0}	0.32	t_P	0.2	x_P	−2.007
y_P^S	0.228	y_P^P	−0.228	k_2^S	−0.1075	k_1^S	−0.3507
k_0^S	0.3329	k_2^P	−0.1295	k_1^P	−0.3269	k_0^P	0.3307
A_R	0.013	ε_R	0.9319	w_{R_0}	0.3663	C_R	0.114
B_R	0.114	Λ_G	1.0	x_R	−2.073	y_R^S	0.22
y_R^P	−0.22	t_R	0.2106	a_H	0.3831	x'_H	−0.3827
γ_R	0.4603	η_R	0.9649	ℓ'_R	−0.9357	κ_R	0.426

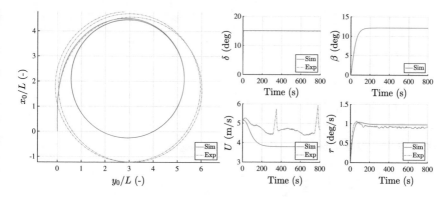

Fig. 6.25 Validation of the 3500 t tanker in the 15° turning maneuver

6.8.3.1 Validation of the Inland 3500 t Tanker in Turning Maneuvers

Simulations of the 3500 t tanker are carried out in 15°, 25°, and 35° turning maneuvers on both starboard and port sides, which are presented in Figs. 6.25, 6.26, 6.27, 6.28, 6.29 and 6.30 respectively. The non-dimensional turning criteria of the simulated and tested results are compared in Table 6.17. The prediction of the 35° and −35° turning maneuvers are quite accurate while those of the 15°, −15°, 25°, and −25° turning maneuvers are not that satisfactory. All the initial turning stages (the first quarters of the turning circles) are well predicted. Therefore, the inaccuracy is primarily caused by the divergence of the steady turn.

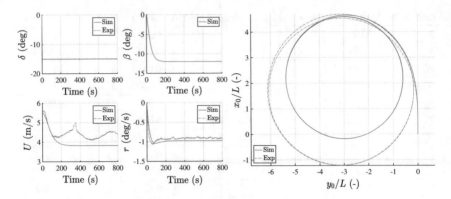

Fig. 6.26 Validation of the 3500 t tanker in the −15° turning maneuver

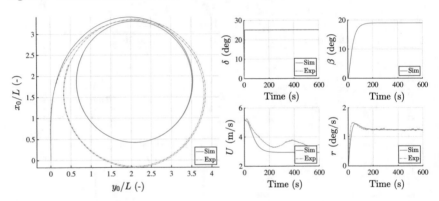

Fig. 6.27 Validation of the 3500 t tanker in the 25° turning maneuver

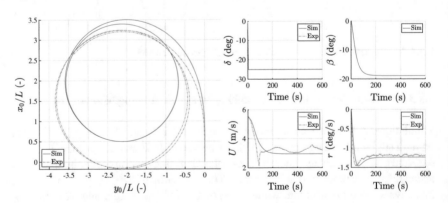

Fig. 6.28 Validation of the 3500 t tanker in the −25° turning maneuver

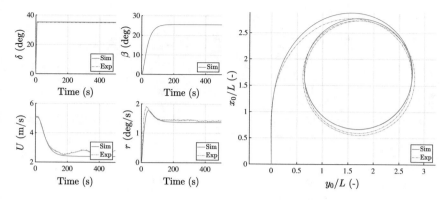

Fig. 6.29   Validation of the 3500 t tanker in the 35° turning maneuver

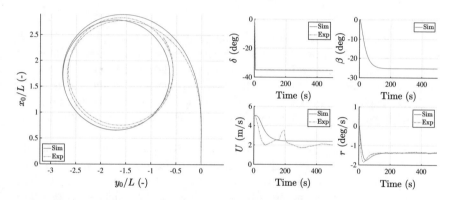

Fig. 6.30   Validation of the 3500 t tanker in the –35° turning maneuver

Table 6.17   Comparison of the simulated and the tested turning indices of the 3500 t tanker

Turning criteria	Sim (–)	Exp (–)	Δ (%)	Turning criteria	Sim (–)	Exp (–)	Δ (%)
A'_D ($\delta = 15°$)	4.46	4.47	−0.22	A'_D ($\delta = -15°$)	4.60	4.43	3.84
T'_D ($\delta = 15°$)	5.23	5.96	−12.25	T'_D ($\delta = -15°$)	5.33	5.97	−10.72
A'_D ($\delta = 25°$)	3.33	3.20	4.06	A'_D ($\delta = -25°$)	3.44	4.01	−14.21
T'_D ($\delta = 25°$)	3.46	3.69	−6.23	T'_D ($\delta = -25°$)	3.51	3.75	−6.40
A'_D ($\delta = 35°$)	2.79	2.63	6.08	A'_D ($\delta = -35°$)	2.80	2.63	6.46
T'_D ($\delta = 35°$)	2.65	2.61	1.53	T'_D ($\delta = -35°$)	2.67	2.64	1.14

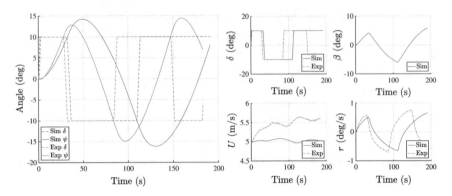

Fig. 6.31 Validation of the 3500 t tanker in the 10°/10° zigzag maneuver

Regarding the constant turning stage, the prediction of the T_D' in the 15°, –15°, 25°, and –25° turning maneuvers is inaccurate while the prediction of T_D' in the 35° and –35° turning maneuvers is accurate. It shows the defect of the hull force module to estimate the forces and moments when the 3500 t tanker is turning with relatively small drift angles. The drift angles for 15°, 25°, and 35° on starboard and port sides turning are about 12°, 18°, 25° respectively. The 3500 t tanker is fitted with a bulbous bow.

When the ship is turning with a small drift angle, the bulbous bow may significantly influence the flow separation at the bow and further affect the effective rudder angle and the rudder inflow speed. However, most of the previous studies on the bulbous bow focused on its effect on powering and few of them considered its impact on maneuvering. The impacts of the bulbous bow on maneuvering forces and moments are not well considered in the presented model and further research is suggested.

6.8.3.2 Validation of the Inland 3500 t Tanker in Zigzag Maneuvers

The heading angles and rudder angles of the 3500 t tanker in 10°/10°, –10°/–10°, 15°/15°, –15°/–15°, 20°/20°, and –20°/–20° are presented in Figs. 6.31, 6.32, 6.33, 6.34, 6.35 and 6.36 respectively. The inaccurate prediction of the speed leads to delays in the time when maximum heading angles are reached.

Table 6.18 compares the simulated and tested overshoot angles. Significant speed drops are observed. Similar to the turning maneuvers, a primary reason for these differences is because the bow effect on the hull forces and moments is not well estimated. The overestimation of the overshoot angles is possibly caused by the fact that the hull damping forces are underestimated. The large predicted speed drop decreases the rudder inflow velocity and reduces the yaw checking force induced by the rudder.

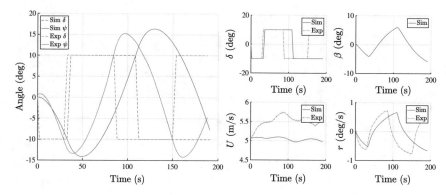

Fig. 6.32 Validation of the 3500 t tanker in the –10°/–10° zigzag maneuver

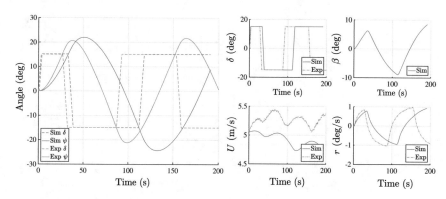

Fig. 6.33 Validation of the 3500 t tanker in the 15°/15° zigzag maneuver

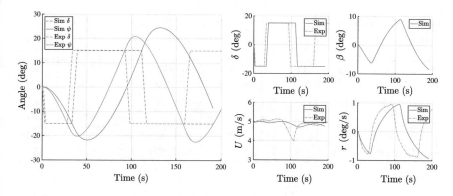

Fig. 6.34 Validation of the 3500 t tanker in the –15°/–15° zigzag maneuver

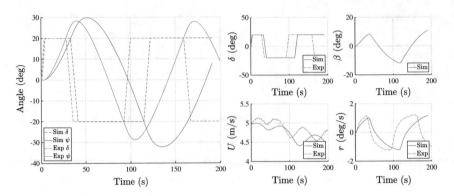

Fig. 6.35 Validation of the 3500 t tanker in the 20°/20° zigzag maneuver

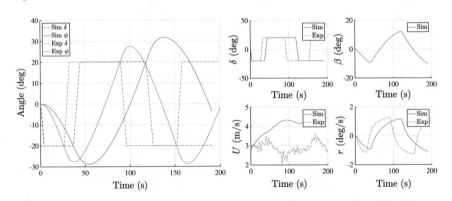

Fig. 6.36 Validation of the 3500 t tanker in the −20°/−20° zigzag maneuver

Table 6.18 Comparison of the simulated and the tested overshoot angles of the 3500 t tanker

Zigzag criteria	Sim (deg)	Exp (deg)	Δ (%)	Zigzag criteria	Sim (deg)	Exp (deg)	Δ (%)
ψ_{O1} (10°/10°)	3.82	2.86	33.57	ψ_{O1} (−10°/−10°)	3.73	3.42	9.06
ψ_{O2} (10°/10°)	5.09	4.95	2.83	ψ_{O2} (−10°/−10°)	5.25	5.28	−0.57
ψ_{O1} (15°/15°)	6.40	5.64	13.48	ψ_{O1} (−15°/−15°)	6.23	5.24	18.89
ψ_{O2} (15°/15°)	7.96	6.18	28.80	ψ_{O2} (−15°/−15°)	8.06	5.92	36.15
ψ_{O1} (20°/20°)	9.27	8.34	11.15	ψ_{O1} (−20°/−20°)	8.21	8.34	−1.56
ψ_{O2} (20°/20°)	10.69	8.81	21.34	ψ_{O2} (−20°/−20°)	10.30	8.81	16.91

6.9 Summary

6.9.1 Concluding Remarks of Inland Vessel Modeling

Ship maneuverability deserves more attention in ship design compared to the previous economic point of view [79], especially for inland vessels. The research on ship maneuverability is increasing, but still mainly focusing on seagoing ships rather than inland vessels. The main challenges for inland vessel maneuverability analysis are how to estimate the hydrodynamic forces with given ship particulars in a specified waterway and how to evaluate their performance according to maneuverability criteria required by the navigation environment of that waterway. Gaps in knowledge need to be filled for inland vessels at slow speed in shallow water considering complex configurations, especially multiple rudders.

An integrated maneuvering model with empirical methods and RANS results has been developed for inland vessels through literature surveys, CFD studies, and analysis of combinations of part models. The presented model uses empirical methods that are publicly available to estimate maneuvering parameters while further improvements can be obtained by carrying out model tests or CFD tests. The regression methods for the hull forces and moment are selected by comparing the simulated and tested maneuvering indices in various maneuvers. Using 2D open water RANS results, the rudder profile, the spacing between rudders, and the endplate effects on the rudder forces and moments can be considered. Furthermore, the proposed modular type maneuvering model allows further research to put new impacts into consideration, such as shallow water, ship-ship, and ship-bank effects.

Maneuvering simulations are performed with one single-propeller single-rudder seagoing KVLCC2 tanker (SPSR S1) and two standard twin-propeller twin-rudder inland vessels (TPTR Y1 and TPTR Y2) from the Yangtze River in various turning and zigzag maneuvers on both port and starboard sides. The simulation results are validated with free-running model tests to show the usability of the proposed model. Through the simulation study, it is confirmed that the presented model can roughly capture the maneuvering characteristics of the reference seagoing and inland vessels. Concluding remarks are drawn as follows:

- The seagoing ship oriented methods are usable for inland vessels while additional attention should be paid to large C_b, L/B, and B/T of inland vessels.
- The asymmetric characteristics are not well captured due to inaccurate input parameters, such as w_P, t_P, w_R, and γ_R. These parameters should be different for each propeller and each rudder on starboard and port side maneuvers.
- The bulbous bow has suspected impacts on the maneuvering performance of the reference 3500 t tanker, especially for zigzag maneuvers. However, these impacts are not well covered in the literature.
- The accuracy of the prediction is quite sensitive to the initial status of the simulation, such as the initial speed and the constant propeller revolution rate. Therefore,

to properly validate the mathematical model, it is necessary to accurately imple-
ment the actual initial states of the validation experiments.

From the perspective of simulations, further investigations and experiments are
requested to describe the inland vessel resistance $(X_H(u))$, express the longitudinal component hull forces due to maneuvering motions $(X_H(v', r')$ or $X_H(\beta, r'))$,
and collect hydrodynamic derivatives for lateral force (Y_H) and yaw moment (N_H).
Moreover, research on the impacts of the bulbous bow on ship maneuverability is
suggested. Additional tests are needed to determine the impact factor of the propeller
(k_P), the rudder (k_R), and the aspect ratio (k_A) on the rudder lift and drag coefficients
during maneuvering motions.

6.9.2 Future Work for Inland Vessel Modeling

To get even more reliable estimates of maneuvering performance of inland vessels,
additional validation data are needed for the applied CFD methods and the mathematical model. Further investigations are also required, which are listed as the research
directions in the following sections. These further research directions are given in
the perspectives of the rudder hydrodynamics, the rudder design, the maneuvering
modeling, shallow-water effects on ship maneuvering, and the interactions among
the hull, the propeller, and the rudder as follows:

- **Rudder hydrodynamics**
  More operational data of different ships with various rudder configurations in
  inland waterways or open sea are needed. Moreover, studies on the rudder performance for ships at slow speed, more specifically at Reynolds numbers lower than
  usual, with uncustomary large rudder angles are suggested.
- **Rudder design**
  Future research is suggested on the investigation of energy-saving methods of the
  rudders such as the twisted rudders and the energy-saving bulbs. Considering cavitation, whole-body profiles and types like the spade rudder with a NACA profile
  have advantages over the separated ones like the semi-skeg rudder with a flap profile. Even though for common commercial ships, the effectiveness and efficiency
  of the rudder get higher priority than the cavitation in the design process, a final
  check of the rudder performance in the cavitation cannot be neglected.
- **Maneuvering modeling**
  Further investigations and experiments are requested to describe the inland vessel
  resistance $(R(u))$, express the longitudinal component hull forces due to maneuvering motions $(R(v', r')$ or $R(\beta, r'))$, and collect hydrodynamic derivatives for
  $(Y_H$ and $N_H)$. Additional tests are needed to determine the impact factor of the
  propeller (ε) and the rudder (ε_R) on the rudder lift and drag coefficients during
  maneuvering motions. Furthermore, the estimation of the wake factors of the propeller (w_P) and the rudder (w_R) needs to be improved.

- **Slow-speed and docking performance**

  This monograph mainly focused on the maneuvering performance of inland vessels at service speed. Due to the limited time and data, slow-speed maneuvering and docking performance were not thoroughly discussed. In slow-speed and docking situations, the hydrodynamic coefficients of the hull, the propeller, and the rudder may be changed. For instance, the rudder will be ineffective owing to the slow inflow to it. The proposed integrated maneuvering model can be improved by adding additional terms to consider the impacts of slow speed, which needs additional tests for conducting the data and validating the model in the future.

- **Shallow water effects on ship maneuvering**

  To predict the maneuvering performance of inland vessels in shallow/confined water, further research is recommended for the estimation of the hull forces and moments. Additionally, the wake factors at the positions of the propeller and the rudder may also be affected by shallow water. Furthermore, it is recommended to study how the flow straightening factor and the effective rudder angle are changed by the variation in water depth.

- **Ship-bank and ship-ship interactions**

  Ship-bank and ship-ship interactions are important for inland vessels. Up to now, these effects are still hard to quantify and, thus, not extensively studied in this monograph. Ship-bank and ship-ship interactions become significant as the distance among them gets smaller. Precise estimation of the start and the amount of the interactions is meaningful to ensure the safety of inland navigation and fully utilize the capacity of inland waterways. Therefore, research on ship-bank and ship-ship interactions of inland vessels is suggested for the future.

- **Interactions among the hull, the propeller, and the rudder**

  As a rule, the interactions among the hull, the propeller, and the rudder affect the rudder hydrodynamic characteristics. The flow straightening effect influences the effective rudder angle, which is important for the calculation of the rudder induced force in maneuvering simulations. However, in the existing literature, the flow straightening factor is primarily determined by model tests. More research is needed to better understand the roles of the impact factors on the flow straightening effect. Furthermore, it is proposed to generate regression formulas of the flow straightening factor through a series of benchmark tests.

  The propeller slipstream mainly delays the stall angle. It may maintain or change the slope of the lift curve depending on the working load of the propeller. Further research in this field would be of great help in maneuvering simulations. The relative position of the propeller and the rudder influence the performance of both the propeller and the rudder. With extra consideration of cavitation, it is recommended to put the rudder as close as possible to the propeller. Moreover, interaction among multiple rudders requires further study, which is particularly meaningful for inland vessels.

References

1. Central Commission for the Navigation of the Rhine (2013) The inland navigation market in 2012
2. Rigo P, Wolters M (2013) In: Rigo P, Wolters M (eds) PIANC - SMART rivers conference 2013, PIANC, Liège, Belgium and Maastricht, The Netherlands
3. International Maritime Organization (2002) Standards for ship manoeuvrability. Resolution MSC.137(76). Accessed 4 Dec 2002
4. International Maritime Organization (2002) Explanatory notes to the standards for ship manoeuvrability. MSC/Circ.1053. Accessed 16 Dec 2012
5. Central Commission for the Navigation of the Rhine (1995) The Rhine vessel inspection regulations
6. European Commission (2008) Commission Directive 2008/126/EC of 19 December 2008 amending Directive 2006/87/EC of the European Parliament and of the Council laying down technical requirements for inland waterway vessels
7. Central Commission for the Navigation of the Rhine (2012) Reglement Onderzoek Schepen op de Rijn (in Dutch)
8. Bureau Veritas (2011) Rules for the classification of inland navigation vessels
9. China Classification Society (2003) Guidelines for inland vessel manoeuvrability (in Chinese). Guidance notes GD-2003
10. ITTC Manoeuvring Committee (2014) Final report and recommendations to the 27th ITTC. In: 27th international towing tank conference, Copenhagen, Denmark, vol 1, pp 128–194
11. Gray WO, Waters J, Blume A, Landsburg AC (2003) Channel design and vessel maneuverability: next steps. Mar Technol 40(2):93–105
12. Landsburg AC, Barr RA, Daggett L, Hwang WY, Jakobsen B, Morris M, Vest L (2005) Critical needs for ship maneuverability: lessons from the Houston ship channel full-scale maneuvering trials. Mar Technol 42(1):11–20
13. Liu J, Hekkenberg R, Rotteveel E (2014) A proposal for standard manoeuvres and parameters for the evaluation of inland ship manoeuvrability. In: European inland waterway navigation conference, Budapest, Hungary
14. Khanfir S, Hasegawa K, Nagarajan V, Shouji K, Lee SK (2011) Manoeuvring characteristics of twin-rudder systems: rudder-hull interaction effect on the manoeuvrability of twin-rudder ships. J Mar Sci Technol 2011(16):472–490
15. Yoshimura Y, Sakurai H (1989) Mathematical model for the manoeuvring ship motion in shallow water (3rd report): manoeuvrability of a twin-propeller twin-rudder ship. J Kansai Soc Nav Archit Jpn 211:115–126
16. Lee H, Kinnas SA, Gu H, Natarajan S (2003) Numerical modeling of rudder sheet cavitation including propeller/rudder interaction and the effects of a tunnel. In: 5th international symposium on cavitation, Osaka, Japan
17. Khanfir S, Hasegawa K, Lee SK, Jang TS, Lee JH, Cheon SJ (2008) Mathematical model for maneuverability and estimation of hydrodynamic coefficients of twin-propeller twin-rudder ship. In: The Japan society of naval architects and ocean engineers, Osaka, Japan, pp 57–60, 2008K-G4-3
18. Gim OS (2013) Assessment of flow characteristics around twin rudder with various gaps using PIV analysis in uniform flow. Ocean Eng 66:1–11
19. Hamamoto M, Enomoto T (1997) Maneuvering performance of a ship with VecTwin rudder system. J Soc Nav Archit Jpn 197–204
20. Lee SK, Fujino M (2003) Assessment of a mathematical model for the manoeuvring motion of a twin-propeller twin-rudder ship. Int Shipbuild Prog 50(1–2):109–123
21. Hasegawa K, Kang D, Sano M, Nabeshima K (2006) Study on the maneuverability of a large vessel installed with a mariner type super VecTwin rudder. J Mar Sci Technol 11(2):88–99
22. Di Mascio A, Dubbioso G, Notaro C, Viviani M (2011) Investigation of twin-screw naval ships maneuverability behavior. J Ship Res 55(4):221–248

23. Nakatake K, Ando J, Kataoka K, Sato T, Yamaguchi K (1989) Study on the propulsive performance of twin screw ship: interaction between propeller and rudder in a uniform flow (in Japanese). Trans West-Jpn Soc Nav Archit 78:49–57

24. Molland AF, Turnock SR (2002) Flow straightening effects on a ship rudder due to upstream propeller and hull. Int Shipbuild Prog 49(3):195–214

25. Kang D, Nagarajan V, Gonno Y, Uematsu Y, Hasegawa K, Shin SC (2011) Installing single-propeller twin-rudder system with less asymmetric maneuvering motions. Ocean Eng 38:1184–1196

26. Coraddu A, Dubbioso G, Mauro S, Viviani M (2013) Analysis of twin screw ships' asymmetric propeller behaviour by means of free running model tests. Ocean Eng 68:47–64

27. Dubbioso G, Mauro S, Ortolani F (2015) Experimental and numerical investigation of asymmetrical behaviour of rudder/propeller for twin screw Ships. In: International conference on marine simulation and ship maneuverability (MARSIM'15), Newcastle upon Tyne, UK

28. Kang D, Nagarajan V, Hasegawa K, Sano M (2008) Mathematical model of single-propeller twin-rudder ship. J Mar Sci Technol 13(3):207–222

29. Eloot K, Vantorre M (2011) Ship behaviour in shallow and confined water: an overview of hydrodynamic effects through EFD. Assessment of stability and control prediction methods for NATO air and sea vehicles, pp 1–20, rTO-MP-AVT-189

30. Quadvlieg F (2013) Theoretische Berekening van Simulatiemodellen voor Binnenvaartschepen ten Behoeve van Maatgevende Manoeuvres (in Dutch). Technical report, Maritime Research Institute Netherlands (MARIN), Wageningen, The Netherlands

31. Yasukawa H, Yoshimura Y (2015) Introduction of MMG standard method for ship maneuvering predictions. J Mar Sci Technol 20(1):37–52. https://doi.org/10.1007/s00773-014-0293-y

32. Kang D, Hasegawa K (2007) Prediction method of hydrodynamic forces acting on the hull of a blunt-body ship in the even keel condition. J Mar Sci Technol 12(1):1–14

33. Kim YG, Kim SY, Kim HT, Lee SW, Yu BS (2007) Prediction of the maneuverability of a large container ship with twin propellers and twin rudders. J Mar Sci Technol 12(3):130–138

34. Bonci M, Viviani M, Broglia R, Dubbioso G (2015) Method for estimating parameters of practical ship manoeuvring models based on the combination of RANSE computations and system identification. Appl Ocean Res 52:274–294

35. Motora S (1959) On the measurement of added mass and added moments of inertia for ship motions (in Japanese). J Soc Nav Archit Jpn 105:83–89

36. Hooft JP, Nienhuis U (1994) The prediction of the ship's manoeuvrability in the design stage. SNAME Trans 102:419–445

37. Motora S (1960) On the measurement of added mass and added moments of inertia for ship motions, Part 2: added mass for the longitudinal motions (in Japanese). J Soc Nav Archit Jpn 106:59–62

38. Motora S (1960) On the measurement of added mass and added moments of inertia for ship motions, Part 3: added mass for the transverse motions (in Japanese). J Soc Nav Archit Jpn 106:63–68

39. Zhou Z, Yan S, Feng W (1983) Manoeuvring prediction of multiple-purpose cargo ships (in Chinese). Ship Eng 6:21–36

40. Clarke D, Gedling P, Hine G (1983) Application of manoeuvring criteria in hull design using linear theory. Trans R Inst Nav Archit 125:45–68

41. Toxopeus S (2011) Practical application of viscous-flow calculations for the simulation of manoeuvring ships. PhD thesis, Delft University of Technology

42. Kulczyk J (1995) Propeller-hull interaction in inland navigation vessel. Trans Built Environ 11:73–89

43. Harvald SA (1983) Resistance and propulsion of ships. Wiley, New York

44. Kristensen HO, Lützen M (2012) Prediction of resistance and propulsion power of ships. Technical report, University of Southern Denmark and Technical University of Denmark, Denmark

45. Hirano M (1980) A practical calculation method of ship maneuvering motion at initial design stage. Nav Archit Ocean Eng 147:68–80

46. Matsumoto K, Sueteru K (1980) The prediction of manoeuvring performances by captive model tests. J Kansai Soc Nav Archit 11–22
47. Hollenbach KU (1999) Estimating resistance and propulsion for single-screw and twin-screw ships in the preliminary design. In: 10th international conference on computer applications in shipbuilding (ICCAS), Cambridge, Massachusetts, USA, vol 2, pp 237–250
48. Holtrop J, Mennen GGJ (1978) A statistical power prediction method. Int Shipbuild Prog 25:253–256
49. Holtrop J, Mennen GGJ (1982) An approximate power prediction method. Int Shipbuild Prog 29(335):166–170
50. Yoshimura Y, Ma N (2003) Manoeuvring prediction of fishing vessels. In: International conference on marine simulation and ship maneuverability (MARSIM'03), Kanazawa, Japan
51. Yoshimura Y, Masumoto Y (2012) Hydrodynamic database and maneuvering prediction method with medium high-speed merchant ships and fishing vessels. In: International conference on marine simulation and ship maneuverability (MARSIM'12), Singapore
52. Nagarajan V, Kang DH, Hasegawa K, Nabeshima K (2008) Comparison of the mariner Schilling rudder and the mariner rudder for VLCCs in strong winds. J Mar Sci Technol 13:24–39
53. Molland AF, Turnock SR (2007) Marine rudders and control surfaces: principles, data, design and applications, 1st edn. Elsevier, Butterworth-Heinemann, Oxford
54. Fujii H (1960) Experimental researches on rudder performance (1) (in Japanese). J Zosen Kiokai 107:105–111
55. Molland AF, Turnock SR (1991) Wind tunnel investigation of the influence of propeller loading on ship rudder performance. Technical report, University of Southampton, Southampton, UK
56. Molland AF, Turnock SR (1992) Further wind tunnel investigation of the influence of propeller loading on ship rudder performance. Technical report, University of Southampton, Southampton, UK
57. Oppenheim BW (1974) A theoretical and experimental investigation of the performance of flapped rudders. Master thesis, Massachusetts Institute of Technology
58. Kerwin JE, Lewis SD, Oppenheim BW (1974) Experiments on rudders with small flaps in free-stream and behind a propeller. Technical report, Massachusets Institue of Technology, Cambridge, MA, USA
59. Nienhuis U (1987) Passieve Manoeuvreerhulpmiddelen: Open Water Proeven met Roer (in Dutch). Technical report, Maritime Research Institute Netherlands (MARIN), Wageningen, The Netherlands
60. Kijima K, Katsuno T, Nakiri Y, Furukawa Y (1990) On the manoeuvring performance of a ship with the parameter of loading condition. J Soc Nav Archit Jpn 168:141–148
61. Furukawa Y, Nakiri Y, Kijima K, Hiroshi I (2008) The prediction of the manoeuvrability of KVLCC1 and KVLCC2. In: SIMMAN 2008: workshop on verification and validation of ship manoeuvring simulation methods, Copenhagen, Denmark, pp 9–14
62. Ankudinov V, Kaplan P, Jacobsen BK (1993) Assessment and principal structure of the modular mathematical model for ship maneuverability prediction and real-time maneuvering simulations. In: International conference on marine simulation and ship manoeuvrability (MARSIM'93), St. John's, Newfoundland, Canada
63. Koh KK, Yasukawa H (2012) Comparison study of a pusher-barge system in shallow water, medium shallow water and deep water conditions. Ocean Eng 46:9–17
64. Yoshimura Y (1986) Mathematical model for the manoeuvring ship motion in shallow water (in Japanese). J Kansai Soc Nav Archit Jpn 41–51
65. Kose K, Yumuro A, Yoshimura Y (1981) Concrete of mathematical model for ship manoeuvring (in Japanese). In: 3rd symposium on ship manoeuvrability, society of naval architects, pp 27–80
66. Eloot K (2006) Selection, experimental determination and evaluation of a mathematical model for ship manoeuvring in shallow water. PhD thesis, Ghent University
67. Lee HY, Shin SS (1998) The prediction of ship's manoeuvring performance in initial design stage. In: Practical design of ships and other floating bodies conference, The Hague, The Netherlands, pp 633–639

68. Broglia R, Dubbioso G, Durante D, Di Mascio A (2013) Simulation of turning circle by CFD: analysis of different propeller models and their effect on manoeuvring prediction. Appl Ocean Res 39:1–10

69. Kim H, Akimoto H, Islam H (2015) Estimation of the hydrodynamic derivatives by RANS simulation of planar motion mechanism test. Ocean Eng 108:129–139

70. Carrica PM, Mofidi A, Eloot K, Delefortrie G (2016) Direct simulation and experimental study of zigzag maneuver of KCS in shallow water. Ocean Eng 112:117–133

71. He S, Kellett P, Yuan Z, Incecik A, Turan O, Boulougouris E (2016) Manoeuvring prediction based on CFD generated derivatives. J Hydrodyn Ser B 28(2):284–292

72. Holtrop J (1984) A statistical re-analysis of resistance and propulsion data. Int Shipbuild Prog 31(363):272–276

73. Pérez FL, Clemente JA (2007) The influence of some ship parameters on manoeuvrability studied at the design stage. Ocean Eng 34:518–525

74. Matsunaga M (1993) Method of predicting ship manoeuvrability in deep and shallow waters as a function of loading condition. Tech Bull Nippon Kaiji Kyokai 11:51–59

75. Ankudinov VK, Jakobsen BK (2006) Physically based maneuvering model for simulations and test evaluations. In: International conference on marine simulation and ship maneuverability (MARSIM'06), Terschelling, The Netherlands

76. Kijima K, Nakiri Y (2004) On the practical prediction method for ship manoeuvrability in restricted water (in Japanese). J Jpn Soc Nav Archit Ocean Eng 37–54

77. Lee SW, Toxopeus SL, Quadvlieg F (2007) Free sailing manoeuvring tests on KVLCC1 and KVLCC2. Technical report, Maritime Research Institute Netherlands (MARIN), Wageningen, The Netherlands

78. Quadvlieg F, Brouwer J (2011) KVLCC2 benchmark data including uncertainty analysis to support manoeuvring predictions. In: 4th international conference on computational methods in marine engineering (MARINE 2011), Lisbon, Portugal

79. Quadvlieg F, Van Coevorden P (2003) Manoeuvring criteria: more than IMO A751 requirements alone! In: International conference on marine simulation and ship maneuverability (MARSIM'03), Kanazawa, Japan, vol 2, pp 1–8

Chapter 7
Tests and Criteria for Inland Vessel Maneuverability

As discussed in Chap. 5, inland vessels are different from seagoing ships in the external environment and internal design factors. These differences should be considered in test maneuvers and related criteria because improper tests may lead to a wrong judgment of the maneuvering performance in reality. The test maneuvers are designed to evaluate the ship's capability of accomplishing certain maneuvering behavior, such as turning in a constrained area, collision avoidance in an emergency, and stopping in a limited distance. Thus, the maneuvers to be designed as practical as possible. However, existing maneuverability standards for inland vessels are not satisfactory. Factors that constrain inland vessel maneuverability such as limited channel breadth and shallow water are not addressed explicitly in the initial ship design. The difficulties of considering the ship behavior in shallow water are commonly ignored on the understanding that if the ship's performance is improved in deep water then it will likely also be better in shallow water.

Standard parameters and maneuvers for design and testing of inland vessel maneuverability are still not as elaborate as the IMO standards for seagoing ships [1, 2]. Thus, guidance and criteria are required to enable naval architects to ensure good maneuverability for navigation safety and economic benefits. For the purpose of achieving more realistic judgment on maneuverability, benchmark maneuvers and related criteria are proposed in this chapter. Section 7.1 introduces the existing maneuverability standards. Accordingly, the contemporary test maneuvers for seagoing ships and inland vessels are presented in Sect. 7.2. After analyzing the definitions in existing standards and maneuvers, Sect. 7.3 proposes new maneuvers and criteria for inland vessels. Some of these maneuvers are tested in Chap. 8. Section 7.4 presents the conclusions of this chapter.

© Springer Nature Switzerland AG 2020
J. Liu, *Mathematical Modeling of Inland Vessel Maneuverability Considering Rudder Hydrodynamics*, https://doi.org/10.1007/978-3-030-47475-1_7

7.1   Existing Maneuverability Standards

Presently, the most common standards for ship maneuverability are issued by International Maritime Organization (IMO). These standards should be applied to "ships of all rudder and propulsion types, of 100 m in length and over, and chemical tankers and gas carriers regardless of the length" [1, 2]. In the IMO standards, the ship type, either seagoing or inland, is not clearly stated. However, the applicable condition of the standards is described as deep, unconstrained water ($H/T > 4$). It is, therefore, rational to suppose that the IMO standards are intended for seagoing ships, but also valuable as guidance for inland vessels.

Regulations for inland vessels are normally proposed by regional authorities [3–5] and classification societies [6]. Compared to the IMO standards for seagoing ships, these regional requirements have fewer test maneuvers and criteria. According to Gray et al. [7], it is still doubtful if the existing standards lead to adequate maneuverability in shallow, restricted, and congested waterways. Therefore, in either case of inland vessels or seagoing ships, new test maneuvers and procedures for shallow and restricted water operations are required to properly predict and evaluate ship maneuverability [8, 9].

In order to improve maritime safety and enhance marine environmental protection, standards for ship maneuverability should be used in ship design, construction, and operation [10]. Due to the lack of uniform maneuvering standards, some ships have been built with very poor maneuvering qualities, which may result in casualties and pollution [11, 12]. Hence, elaborate and uniform criteria should be established for safety [13]. ITTC Manoeuvring Committee [14] gave a review of the criteria in use for inland vessels, fast ships, and dedicated low-speed maneuvers.

More requirements are necessary for other scenarios like constrained waterways and port areas. The necessity of more critical requirements for specific situations was discussed by Li et al. [15]. Table 7.1 compares the existing maneuverability standards. From this table, it is concluded that more elaborate criteria for different navigational conditions, especially for inland waterways, should be issued to define the minimum performance. Furthermore, emergency situations, such as engine failure, strong wind, currents, and waves, should be examined to predict the worst maneuvering cases.

Due to the differences in navigation conditions and ship particulars between inland vessels and seagoing ships as discussed in Sect. 5.1, the standards for maneuverability are expected to be 'different' in the aspects of the test maneuvers and criteria. A summary of existing standards, namely International Maritime Organization (IMO), American Bureau of Shipping (ABS), European Commission (EC), Central Commission for the Navigation of the Rhine (CCNR), and Bureau Veritas (BV) is given in the succeeding paragraphs to find the gaps to improve inland vessel standards. After discussing the existing maneuverability standards, the required full-scale test conditions and contents are proposed.

Table 7.1 Overview of existing standards and criteria for ship maneuverability [1, 2, 4, 5, 16]

Abilities	Manoeuvres	IMO	ABS	CCNR and European Commission	Bureau Veritas
Turning	35° turning	Advance < 4.5 L. Tactical diameter < 5 L.	Not rated R_{td} >= 1	The turning capacity of vessels and convoys whose length does not exceed 86 m and width does not exceed 22.90 m shall be considered sufficient.	
Initial turning	10°/10° zigzag	Distance ship travelled <= 2.5 L by the time the heading has changed by 10° from the original heading	R_{ti} >= 1		
Yaw checking and course keeping	10°/10° zigzag	First overshoot angle: < 10° (L/V < 10 s); < (5+0.5 L/V)° (10 s ≤ L/V < 30 s); < 20° (L/V ≥ 30 s).	Rated >= 1	Evasive manoeuvres with a rudder angle of 20° and 45° to starboard and port shall be checked by yaw rate and maximal period instead of overshoot angles for zig-zag manoeuvres. Criteria vary for different ship dimensions and water depth.	Requirements are put on stopping capacity, astern trials, capacity of taking evasive actions, and turning capacity of inland vessels.
		Second overshoot angle: < 25° (L/V < 10 s); < (17.5 + 0.75 L/V)° (10 s ≤ L/V <30 s); < 40° (L/V ≥ 30 s).	Not rated		
	20°/20° zigzag	First overshoot angle ≤ 25°	R_{fa20} >= 1		
Stopping	Full astern stopping	Track reach < 15 L. None for head reach.	Not rated R_{ts} >= 1	In flowing or standing water, stopping distance changes with the ship length.	

7.1.1 International Maritime Organization

The most widely accepted criteria for ship maneuverability are issued by International Maritime Organization (IMO), including turning ability, initial turning ability, yaw checking and course keeping ability and stopping ability [1, 2]. Daidola et al. [17] described how these IMO standards were defined and improved. However, these standards are specified for ships longer than 100 m in deep unconstrained water. Shorter ships and vessels with unconventional propulsion systems, for instance, azimuth thrusters, are not subject to the IMO standards. Based on the opinion of the administration, current rules can be taken as reference for unconventional ships.

To comply with the requirements of authorities, the maneuverability criteria should be evaluated under specified test conditions and procedures [18]. For seagoing ships, the trial should be conducted in deep unconstrained sea water to eliminate the effects of the waterway bottom, banks, and other external objects. The deep water here means that the depth of water should be more than 4 times the mean draft [2]. The trial speed should be set to at least 90% of the ship's speed corresponding to 85% of the maximum engine output. The test ships should be loaded to the design dead weight and even keel within 5% deviation.

Three maneuvers are needed for sea trials: the turning circle maneuver (turning and initial turning ability), the zigzag maneuver (yaw checking and course keeping abilities), and the stopping test (stopping ability) [1, 2]. For example, the zigzag test, which is specially developed for towing tank tests but also popular for full-scale tests [19], can show the maneuvering capacities of initial turning and yaw checking ability. Brix [20] carried out a series of model tests, yielding the typical values of ship zigzag indices. Since the test environment described in the IMO standards is open deep water, there is a need to consider ship's maneuvering capacities in shallow constrained areas, such as harbor entrance channels and ports [7, 8, 21].

The existing maneuvers may also lead to a misunderstanding of the actual maneuvering performance. Yoshimura et al. [22] pointed out that an evaluation based on the second overshoot angle of a 10°/10° zigzag test and the first overshoot angle of a 20°/20° zigzag test may regard poor maneuverability as good. At present, hydrodynamic (constrained water), meteorological (the wind, the wave, and the current), and navigational (other ships, artificial constructions) impacts are not covered in the existing IMO standards. The need to formulate criteria for off service speed (slow speed) and water depth (shallow water) was recognized by Dand [23], Gray et al. [7], Hwang et al. [21], Landsburg et al. [8], Quadvlieg and Van Coevorden [24].

Besides the standards specified for ship maneuverability, there are other requirements of IMO for navigation safety, which indirectly affect the ship maneuvering performance. For example, International Maritime Organization [25] states, "The main steering gear and rudder stock should be capable of putting the rudder over from 35° on one side to 35° on the other side with the ship at its deepest seagoing draft and running ahead at maximum ahead service speed and, under the same conditions, from 35° on either side to 30° on the other side is not more than 28 s". This rule requires a minimal rudder turning rate, which affects the turning related maneuvering abilities.

7.1.2 American Bureau of Shipping

In most of the cases, ship designers only want to meet the minimum requirements of the authorities [24]. In order to enhance navigation safety instead of just meeting the minimum criteria, American Bureau of Shipping (ABS) built a rating system to evaluate the overall maneuvering capacity, which provides information on implementation and application of the IMO standards [16]. Biancardi [10] developed a set of performance indices based on full-scale trial results. Spyrou [26] applied a rating procedure based on a synthesized maneuverability index. Belenky and Falzarano [27] compared the IMO requirements [1, 2], the rating system established by Barr et al. [28], and ABS Guide for Vessel maneuverability [16].

7.1.3 Central Commission for the Navigation of the Rhine

Even though there are no universal criteria such as IMO standards for inland vessels, requirements have been set up as regional regulations. To maintain a safe and smooth flow of traffic on the Rhine, maneuvering criteria and assessment approaches were proposed by Dijkhuis et al. [12]. Central Commission for the Navigation of the Rhine (CCNR) stated the required inland vessel maneuverability in terms of the forward speed, the stopping capacity, the maneuverability while going astern, the capacity to take evasive actions, and the turning capacity [3]. Only the evasive maneuver (similar to zigzag tests but checked by yaw rates instead of heading angles) and the stopping test are mandatory [5]. However, as seen in Table 7.1, these criteria are not as elaborate as the IMO standards. Since inland vessels commonly sail in more complex situations than seagoing ships, they may need more test maneuvers and criteria in detail to ensure navigation safety.

Considering the differences between the open sea and inland waterways, inland trials should be carried out in representative inland waterways. For ships in the Rhine, tests should be performed in areas of the Rhine or other inland waterways with similar conditions [5]. The area should be straight, not less than 2 km, and sufficiently wide, in flowing or standing water. The under keel clearance should be at least 20% of the water depth and not less than 0.5 m [3]. The test load condition should be 70% to 100% of full load and even keel. The ship velocity relative to the water is at least 13 km. Vessels and convoys proceeding downstream should be able to stop in good time while remaining sufficiently maneuverable. Turning capacity should be demonstrated by upstream turning maneuvers.

7.1.4 European Commission

European Commission (EC) states that the technical requirements related to the capacity of taking evasive action and turning [4]. For changes in speed, it concerns the forward speed, the stopping capacity, and the capacity of going astern. Accordingly, the maneuvering capacities are tested by evasive action test and full astern test [4, 5]. Furthermore, these requirements are also applied to coupling systems in a rigid assembly and vessels with active steering devices. At least four evasive action maneuvers are requested, i.e. port and starboard side tests with rudder angles of 20° and 45°. The test conditions are similar to those of the CCNR standards in Sect. 7.1.3.

7.1.5 Bureau Veritas

Bureau Veritas (BV) put requirements of inland vessel maneuverability on the stopping capacity, the astern trials, the capacity of taking evasive actions, and the turning ability of inland vessels [6]. The ships classified by Bureau Veritas should be checked by all navigation tests in areas designated by the ship classification society. Similar to the European Commission standards, the stopping capacity test for ships which are not longer than 86 m and not wider than 22.9 m can be replaced by turning maneuvers [6]. The requirements on ship maneuverability in the Bureau Veritas are less elaborate than other standards as their focus is mainly on the structure and the equipment of the ship.

7.1.6 Concluding Remarks of Existing Maneuverability Maneuvers

After reviewing the existing standards, it is clear that the expected differences between regulations for inland vessels and seagoing ships exist, but the requirements have not always been specified in detail. Thus, based on the existing standards, new maneuvers and parameters are needed to have a deeper insight into the maneuvering performance of inland vessels. Furthermore, the existing methods of the full-scale test are also insufficient for ships in shallow/constrained water. In order to help naval architects to evaluate inland vessel maneuvering performance, more elaborate criteria are needed. Since navigation conditions and ship particulars are different for inland vessels and seagoing ships, evaluation maneuvers and related criteria should be adjusted. The existing test maneuvers are reviewed in Sect. 7.2. Additionally, new test maneuvers are proposed in Sect. 7.3.

7.2 Existing Test Maneuvers

The main contemporary test maneuvers for seagoing ships and inland vessels are regulated by Central Commission for the Navigation of the Rhine [5], European Commission [4], International Maritime Organization [1, 2]. For seagoing ships, the test maneuvers are turning circle, zigzag, and stopping [1, 2]. In addition, International Maritime Organization [1, 2] suggests additional maneuvers, including spiral, reverse spiral, simplified spiral, pull-out, and very small zigzag maneuvers, to further investigate the dynamic stability characteristics of the ship. For inland vessels, the test maneuvers are evasive action and stopping [4, 5]. The test conditions have been described in Sect. 7.1. The following paragraphs present these test maneuvers and discuss the insufficiency of applying them to inland maneuverability test.

7.2.1 Existing Turning Circle Test

As requested by International Maritime Organization [1, 2], the turning circle maneuver should be performed on both starboard and port sides with 35° or the maximum design rudder angle permissible at the test speed when the ship is approaching at zero yaw rate. International Towing Tank Conference [29] indicated that a turning circle of at least 540° is necessary to determine the main turning indices while International Maritime Organization [2] recommended that a 720° turn should be completed to fully assess the environmental effects. Figure 7.1 shows the terminologies used on the turning circle test.

International Maritime Organization [2] lists the tactical diameter, the advance, and the transfer as the essential parameters during a turning test. Besides these three

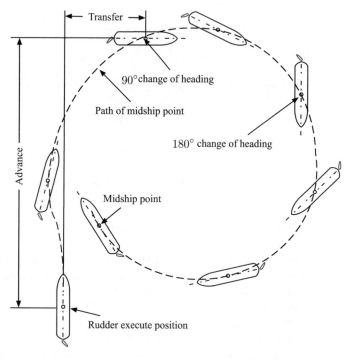

Fig. 7.1 Terminologies used on existing turning circle tests. Adapted from International Maritime Organization [2]

parameters, International Towing Tank Conference [29] suggested to additionally obtain the loss of speed in a steady turn, the time to change heading 90° and 180°, the maximum advance, and the maximum transfer. However, International Maritime Organization [1] only takes the tactical diameter and advance as criteria for turning ability. It should be noted that there is no turning circle maneuver or criterion specified for inland vessels.

7.2.2   Existing Zigzag Test

The zigzag maneuver is performed to evaluate the yaw checking and course keeping abilities. Figure 7.2 illustrates the definition used on 20° (the heading angle ψ)/20° (the rudder angle δ) zigzag test. International Maritime Organization [1, 2] requests 10°/10° and 20°/20° zigzag test with both starboard and port rudder angles to identify the environmental effects. Moreover, International Towing Tank Conference [29] indicated that the turning and the yaw tracking abilities are of special interest as the emergency turns should be carried out to starboard. Values of first and second

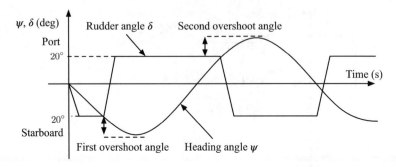

Fig. 7.2 Terminologies used on existing 20°/20° zigzag tests. Adapted from [2]

overshoot angles in 10°/10° zigzag are given while only the limit of the first overshoot angle in 20°/20° is set [1, 2].

Besides the overshoot angles, International Towing Tank Conference [29] defines additional results of the zigzag test, including the initial turning time, the time to check yaw, the reach, the time of a complete cycle, the angular speed, and the unit time. To express the course keeping ability in conditions similar to practice, modified zigzag [29] and very small zigzag [2] were proposed, which may have the execute heading angle as small as 1° with the rudder angle being 5° or 10°. Nevertheless, the zigzag test might not be sufficient for ships in constrained waterways, especially for the case of inland vessels. For instance, the maximum deviation from the original course during the zigzag maneuvering is valuable to ensure that inland vessels do not hit the bank in emergency collision avoidance.

7.2.3 Existing Evasive Action Test

The evasive action test is used to prove that the tested inland vessel can take emergency avoidance in good time [4, 5]. Figure 7.3 presents the terminologies used on the evasive action test. The procedure of the evasive action test is similar to the zigzag test, but the criteria are different. The evasive action test evaluates the required yaw rate (r_1 and r_3) and the time to reach the second zero yaw rate (t_4) instead of the overshoot angles. r_1 and r_3 are different for the various dimensions of vessels or convoys. Furthermore, the limit value of t_4 depends on the ratio of the water depth to the ship draft (H/T). More information is achievable from this test, such as the reach from t_0 to t_4, the maximum deviation from the original course, and the deviations from the original course when r_1 and r_3 are achieved, but these values are not evaluated in the contemporary standards.

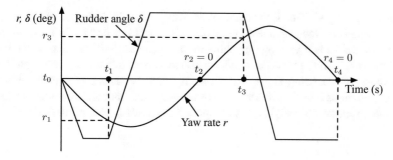

Fig. 7.3 Terminologies used on existing evasive action tests. Adapted from [4]

7.2.4 Existing Stopping Test

Stopping ability is tested by the full astern stopping. The test procedure described by Central Commission for the Navigation of the Rhine [5], European Commission [4], International Maritime Organization [1, 2] are basically the same. Figure 7.4 shows the terminologies used on the stopping test. It should be noted that the rudder angle is requested to be maintained at a neutral position during the trial. In these regulations, only the track reach, which is the distance (relative to the ground) traveled by the midship, is checked. Rules for inland vessels [4, 5] consider the current (flowing or standing water) and the ship dimensions (length and width) while those for seagoing ships only mention that the track reach may be modified by the administration.

Ships can be unstable in direction during the stopping maneuver, so that the trajectory is, to a large extent, determined by the initial conditions and environmental

Fig. 7.4 Terminologies used on existing stopping tests. Adapted from [2]

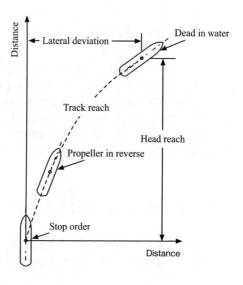

disturbances [2, 29]. That is the main reason why the lateral deviations are of interest but not evaluated. However, these lateral deviations are more crucial for inland vessels in narrow channels than seagoing ships in the open sea. Furthermore, the ambient disturbances, mainly about the wind, on inland vessels are also far less significant than those for seagoing ships. Therefore, it might be meaningful to include the lateral deviations into the criteria. Moreover, instead of maintaining a neutral rudder position, proper rudder angles may be applied to reduce the lateral deviations.

7.2.5 Concluding Remarks of Existing Maneuverability Standards

Section 7.1 has reviewed the existing maneuvers for inland vessels (evasive action and stopping) and seagoing ships (turning, zigzag, and stopping). The deficiency of these maneuvers and the missing criteria has been discussed. The existing test maneuvers set a regulatory framework to address the fundamental maneuvering abilities of ships. However, the existing criteria and maneuvers for inland vessels are not as elaborate as those for seagoing ships. Furthermore, the existing criteria do not present the impacts of ship maneuverability on navigation efficiency.

Due to the differences in sailing conditions, the parameters that are not important for seagoing ships can be crucial for inland vessels, such as the lateral deviations in the zigzag test and the stopping test. Additionally, high-level data can be abstracted from the existing maneuvers, such as the speed drop and the operation time, to show the impacts of maneuvering on daily operations. Following this discussion, new maneuvers and standards for inland vessels are proposed in Sect. 7.3.

7.3 Proposed Test Maneuvers

As discussed in Sect. 7.2, test maneuvers need to be adapted to the actual sailing conditions on restricted water. Furthermore, for both seagoing ships and inland vessels, it is possible to obtain more information from the existing test maneuvers than those are regulated by the current standards. Accordingly, this section proposes test maneuvers and related maneuvering parameters for inland vessels. These maneuvers and parameters are then applied in Chap. 8 and further summarized in Chap. 9.

7.3.1 Proposed Turning Circle Test

As previously mentioned, the turning circle test is not mandatory for inland vessels. Due to the dimensions of the inland waterways, large inland vessels may not make

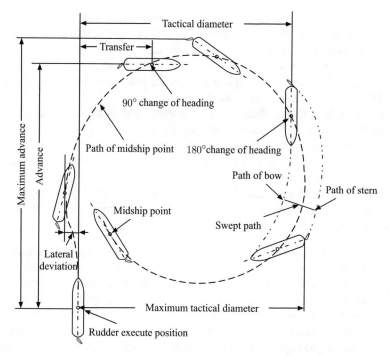

Fig. 7.5 Terminologies used on the proposed turning circle test for inland vessels

the full turning circle at the cruising speed, for instance, 110 m inland vessels in the Rhine. In some circumstances, the turning circle maneuver is possible for small vessels or inland vessels in large rivers like the downstream of the Yangtze River. However, these full turning maneuvers are strictly limited or even not allowed in some area. In practice, inland vessels with bow and stern thrusters can turn on a spot. However, the bow and stern thrusters are only effective at slow speed (under 6 km h^{-1}).

Regarding the limits of the turning circle test for inland vessels, it is still an effective way to evaluate a ship's turning ability. For vessels that sail in relatively narrow waterways where a full turning circle is not possible, the turning circle test is not necessary. Then, the turning ability of these vessels should be checked by other test maneuvers. For inland vessels in large rivers, the turning circle is recommended and new parameters are proposed based on the existing turning circle test (Sect. 7.2.1) as illustrated in Fig. 7.5.

More specifically, the proposed parameters are explained as follows:

- The **advance**, the **transfer**, and the **tactical diameter** are defined as the same as the current standards.

- The **lateral deviation** is the distance from the position at which a starboard rudder angle is given to the maximum port side deviation of the midship point of a vessel in a starboard side turning circle, and vice versa.
- The **maximum advance** is the distance travelled by the stern of a vessel from the position at which the rudder angle is given to the position at which the ship speed in the direction of the original course is zero.
- The **maximum tactical diameter** is the distance travelled by the stern of a ship from the position at which the rudder angle is given to the position at which the heading has changed 180° from the original course.
- The **maximum swept path** is the maximum difference between the trajectories of the bow and the stern.
- The **differences of the tactical diameters and advances** between the starboard side and port side turning circle tests should be studied to present the asymmetry behaviors of the ships.

Even though the classic turning maneuver can give a good representation of ship turning ability, the full circle turning test requires so much space that sometimes it is not feasible to be carried out for operation in real inland navigation. Furthermore, it is not a maneuver that an inland vessel is likely to make. Additionally, inland vessels are more commonly configured with multiple rudders and the maximum rudder angles for inland vessels can be 75° or even 90° instead of the customary 35° for seagoing ships. Therefore, the applied rudder angle should be clearly specified. Otherwise, the non-dimensional turning indices of inland vessels can be quite different from those of seagoing ships. Considering that it is not possible to use the turning circle test in narrow waterways, a hard turning maneuver is proposed in the Sect. 7.3.2.

7.3.2 Proposed Hard Turning Test

Hard turning test is proposed to present the turning ability of ships in bendy waterways. For a hard turning maneuver, the inland vessel needs to alter its course by a large angle with a specific rudder angle and then keep the new course. The extreme case of changing the course by 90° is just the first quarter of a turning circle as shown in Fig. 7.6. More frequently, inland vessels need to change the course by 45° to 60°. The target performance parameters are the lateral deviation, the advance, and the transfer, which are defined as the same as the terminologies defined in Sect. 7.5. Furthermore, the operation time and the speed drop are meaningful to evaluate the ship response to the applied rudder angle. For cases where the turning circle test is not applicable, this hard turning maneuver can be an alternative to evaluate the ship turning ability.

Fig. 7.6 Terminologies used
on proposed hard turning
circle tests for inland vessels

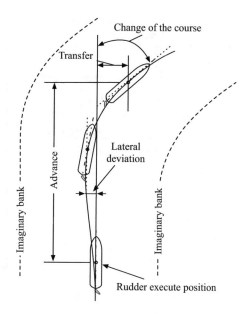

7.3.3 Proposed T-Junction Test

The confluence of rivers, channels, and canals are challenges for inland vessel navigation. Near the confluences, more encountering situations happen, increasing the density of the traffic flow. Besides, the speed and the direction of the current may change sharply, affecting the ship maneuverability. Furthermore, observation of the skippers may be affected by natural or artificial obstacles. All in all, inland vessels should be capable of making a proper motion in the junction area and have sufficient maneuvering margin in case of emergency.

The T-junction maneuver can be taken as hard turning with extra constraints, such as the safe distance from the banks and the remaining yaw rate and the resultant drift angle after the operation. Thus, the parameters used for hard turning tests, such as the advance and the transfer, can also be applied to the T-junction maneuvers. In this section, example trajectories of a large inland vessel or an inland vessel that has poor maneuverability on single-lane and double-lane T-junctions are shown in Fig. 7.7.

According to Rijkswaterstaat [30, p. 39], the width in the plane of the waterway bottom and the width in the keel plane of single-lane waterways must be at least twice the width of the reference ship. Double-lane waterways are divided as normal profile and narrow profile. The width in the plane of the waterway bottom for both double-lane profiles should be at least 2 times the width of the reference ship. The width in the keel plane of the loaded reference vessel must be at least 4 and 3 times the width of the reference vessel in the normal profile and narrow profile respectively [30, p. 39].

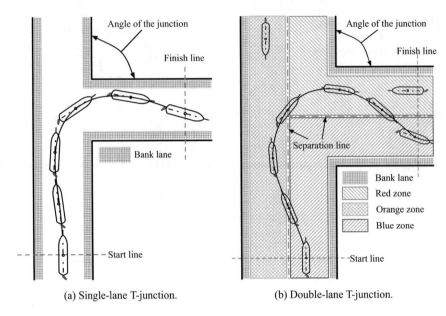

(a) Single-lane T-junction.   (b) Double-lane T-junction.

Fig. 7.7 Terminologies used on proposed T-junction tests for inland vessels

The proposed single-lane T-junction test starts when the midship point of the ship passes the imaginary start line and finishes when the midship point of the ship reaches the finish line. The start and end speeds should be recorded to calculate the speed drop. The duration of the operation is used to show the quickness of the vessel to the turning order. Furthermore, the remaining drift angle and the resultant yaw rate at the endpoint are used to present the required maneuvering margin to turn the vessel back to the straight course and avoid entering the bank lane. In general, vessels with smaller parameters, namely the speed drop, the operation time, the remaining yaw rate, and the resultant drift angle, have better maneuvering performance.

The double-lane T-junction is more complex than the single-lane one because encountering situations may happen while the ship is turning. In some areas, a strict Traffic Separation Scheme (TSS) is issued, where the vessel is not allowed to cross the separation line. Taking the separation line as an imaginary bank, the double-lane is the same as the single lane. To fully use the capacity of the waterway, large vessels may be allowed to cross the separation line with extra care. For waterways where the TSS is not applied, double-lane T-junction is suggested.

The navigable area of the double-lane channel is divided into three zones, i.e. blue, orange, and red. The blue zone is the safe zone for the own vessel. A vessel with good maneuverability is expected to be capable of passing the T-junction in the blue zone. The orange and red zones are lanes for the incoming vessels. Therefore, it is dangerous for the own ship to enter these zones. Also, the red zone is more dangerous than the orange zone because the incoming ship in the red zone needs to make its T-junction and has a relatively small vision due to the corner of the channels.

Similar to the case of the single-lane T-junction, the vessel is not allowed to enter the bank lane. In addition, the safety of the maneuver is judged by calculating the area between the ship trajectory and the separation line in each zone. The larger area in the orange and the red zones, the more dangerous the maneuver is. Furthermore, the maneuvering performance of the vessel is evaluated based on the velocity, the drift angle, and the yaw rate at the time point when the ship enters or leaves each zone. For instance, the operation time in the orange zone and the red zone should be minimized.

To quantify the maneuvering performance of a ship, the previously described turning, hard-turning, lane-changing tests are more fundamental than the proposed T-junction test. The properties of the T-junction tests depend on the waterway that ships sail on. For simplicity, the T-junction turning maneuver can be regarded as turning or hard-turning with additional boundaries. These boundaries should be considered by the naval architects for specified situations. In the rest of the monograph, the T-junction is not discussed explicitly, but the performance of ships in turning and hard-turning can show some insights into the possible behaviors in T-junction tests.

7.3.4 Proposed Lane Changing Test

One of the most important aspects of the inland vessel maneuverability is the capability to change lanes in ship encountering and overtaking [12]. However, the lane changing ability is not covered in the existing standards. The capability of changing lanes is related to the initial turning ability and yaw checking ability, which can roughly be presented by the classic zigzag test. However, zigzag tests may lead to a wrong estimation of ship maneuvering performance. The large overshoot angle may be caused by the large inertia of large ships or by the large rate of turn of small ships.

Dijkhuis et al. [12] proposed a revised zigzag test, which is based on the change of the rate of turn. Nevertheless, it is more realistic to emphasize the capability of a single lane changing test instead of continuous maneuvers like the zigzag test. An example of the proposed lane changing test in an overtaking situation is illustrated in Fig. 7.8. The distance before overtaking, the distance after overtaking, and the lateral distance should be expressed in non-dimensional forms of both ships' properties and parameters of the channel.

Considering the small relative speed in the overtaking operations, it may be less critical than the cases of ship encountering. In more serious situations of collision avoidance, ships have to make large course altering to avoid the obstacles and correct the course as soon as possible to prevent grounding or ship-bank collision. In that case, large angles of lane changing tests are needed, for instance, 35°/35° or even larger lane changing tests. To evaluate the effectiveness of the large lane changing tests, further understanding of the rudder performance is needed.

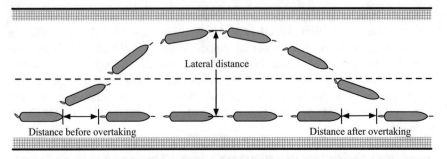

Fig. 7.8 Terminologies used on proposed lane changing tests

7.3.5 Proposed Stopping with Rudder Correction Test

Stopping ability in a straight channel is important for emergency operations. As the channel is constrained in width, the stopping scenario for inland vessels is different from that for seagoing ships. The basic IMO requirement of the stopping ability is that the track reach should be smaller than 15 ship lengths (except impracticable cases for large displacement ships) in the full astern stopping test. To avoid a collision, seagoing ships commonly choose the more efficient operation of hard turning while, due to the constraints of the waterways, inland vessels have to carry out the crash stopping.

In the existing standards, there is no description of the rudder in the stopping test. The rudder angle is commonly taken as zero during the whole operation, more specifically from the time when the full astern order is given till the ship stops in the water. However, for inland ships, both the advance and the lateral deviation should be considered. To reduce the lateral deviation, inland vessels may use the rudder correction force as the proposed maneuver in Fig. 7.9. In this case, there is the question of when to start this correction rudder order and when to stop it.

7.3.6 Proposed Stopping with Clam Shell Angles Test

Twin-rudder inland vessels can set their rudders both outwards to reduce the crash stopping distance. Unlike seagoing ships that normally have a maximum rudder angle of 35° and operate both rudders in the same direction, inland vessels may apply rudder angles of nearly 90° to both sides. These outwards rudder angles are named as clam shell angles as shown in Fig. 7.10.

An increase in the clam shell angle reduces the gap between the leading edges of the twin rudders. Therefore, more resistance, which is good for stopping, is induced

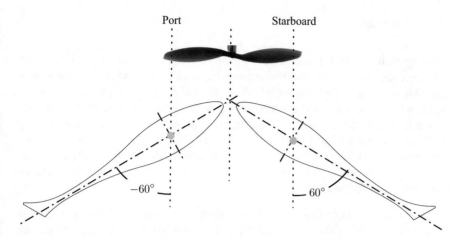

Fig. 7.9 Terminologies used on proposed stopping with rudder correction tests

Fig. 7.10 An example of the rudder clam shell angles

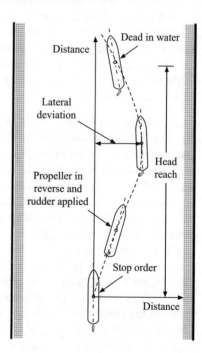

by the rudders. The stopping distance can be reduced by 50% with the clam shell angles [31, p. 41]. In addition to the track reach required in the current standards, both the transfer and the head reach should be recorded as they are critical for inland vessel safety.

7.4 Summary

In Chap. 5, the main impact factors on maneuvering performance were reviewed. Through this analysis, clear differences were found in the navigation environment and ship configurations between inland vessels and seagoing ships. In this chapter, standards of maneuvers and criteria on ship maneuverability have been compared. There is a lack of knowledge of suitable maneuvers to evaluate the maneuvering performance of inland vessels in real-world navigation. Therefore, new benchmark maneuvers have been proposed for further discussion. After all, the practical maneuvers to evaluate and compare the maneuvering performance of inland vessels are summarized as follows:

- Test maneuvers and related criteria for inland vessels are less elaborate than those for seagoing ships. The current test maneuvers have certain insufficiency to present the maneuvering performance of day-to-day operations in inland waterways.
- Besides the maneuvering indices in the current standards, additional parameters can be obtained from the existing test maneuvers. It is possible to have further insight into the ship maneuverability with the classic maneuvering tests.
- Inland vessels have more difficult maneuvering situations like the T-junction than seagoing ships. Furthermore, inland vessels have extraordinary operation profiles like stopping with clam shell angles. These differences require extra care to ensure navigation safety.

Due to the possibilities of nautical operations, interest in maneuvers for seagoing ships and inland vessels is not always the same. For instance, in the case of an imminent collision, a seagoing ship may either initiate a crash stopping or a turning maneuver while inland vessels can only make a crash stopping due to the limits of waterways. This highlights the importance of the crash stopping ability for inland vessels. All these different impact factors and interests in the maneuvering performance request further research on the mechanisms of ship motions for more accurate maneuverability prediction [19].

Based on the existing test maneuvers and practical operations, new parameters and maneuvers have been proposed to evaluate inland vessel maneuverability in this chapter. From a perspective of initial design, classic turning and zigzag tests and proposed hard-turning and lane-changing maneuvers are used to compare the maneuvering performance of inland vessels with various rudder configurations in Chap. 8. The proposed T-junction, stopping with rudder correction, and stopping with clam shell angles can be carried out with specified safety margins to authorities' requirements.

In Chap. 4, hydrodynamic characteristics of rudders with various profiles and parameters are studied through CFD simulations. These hydrodynamic characteristics are then integrated into the maneuvering model built in Chap. 6. With this integrated maneuvering model, inland vessels with different rudder configurations are tested in the proposed test maneuvers.

References

1. International Maritime Organization (2002) Standards for ship manoeuvrability. Resolution MSC.137(76). Accessed 4 Dec 2002
2. International Maritime Organization (2002) Explanatory notes to the standards for ship manoeuvrability. MSC/Circ.1053. Accessed 16 Dec 2012
3. Central Commission for the Navigation of the Rhine (1995) The Rhine Vessel Inspection Regulations
4. European Commission (2008) Commission directive 2008/126/EC of 19 December 2008 amending directive 2006/87/EC of the European parliament and of the council laying down technical requirements for inland waterway vessels
5. Central Commission for the Navigation of the Rhine (2012) Reglement Onderzoek Schepen op de Rijn (in Dutch)
6. Bureau Veritas (2011) Rules for the classification of inland navigation vessels
7. Gray WO, Waters J, Blume A, Landsburg AC (2003) Channel design and vessel maneuverability: next steps. Marine Technol. 40(2):93–105
8. Landsburg AC, Barr RA, Daggett L, Hwang WY, Jakobsen B, Morris M, Vest L (2005) Critical needs for ship maneuverability: lessons from the Houston ship channel full-scale maneuvering trials. Marine Technol. 42(1):11–20
9. Liu J, Hekkenberg R, Rotteveel E (2014) A proposal for standard manoeuvres and parameters for the evaluation of inland ship manoeuvrability. In: European inland waterway navigation conference, Budapest, Hungary
10. Biancardi CG (1993) Integrating ship manoeuvrability with safety. In: International conference on marine simulation and ship maneuverability (MARSIM '93), St. John's, Newfoundland, Canada, vol 1, pp 33–47
11. Nobukawa T, Kato T, Motomura K, Yoshimura Y (1990) Studies on manoeuvrability standards from the viewpoint of marine pilots. In: International conference on marine simulation and ship maneuverability (MARSIM '90), Tokyo, Japan
12. Dijkhuis T, Van Toorenburg JCK, Verkerk F (1993) New manoeuvring criteria for the river Rhine. In: International conference on marine simulation and ship maneuverability (MARSIM '93), St. John's, Newfoundland, Canada, vol 1, pp 77–86
13. Pérez FL, Clemente JA (2007) The influence of some ship parameters on manoeuvrability studied at the design stage. Ocean Eng 34:518–525
14. ITTC Manoeuvring Committee (2014) Final Report and Recommendations to the 27th ITTC. In: 27th International towing tank conference, Copenhagen, Denmark, vol 1, pp 128–194
15. Li Y, Landsburg AC, Barr RA, Calisal SM (2005) Improving ship maneuverability standards as a means for increasing ship controllability and safety. MTS/IEEE OCEANS 2005. DC, USA, Washington, pp 1972–1981
16. American Bureau of Shipping (2006) Guide for vessel maneuverability
17. Daidola JC, Lundy W, Barr R (2002) Evolution of the IMO standards for maneuverability. SNAME Trans 110:395–411
18. International Towing Tank Conference (2008) ITTC recommended procedures and guidelines 7.5-02-03-02.1: Testing and extrapolation methods propulsion, Propulsor Open Water Test
19. Bertram V (2012) Practical ship hydrodynamics, 2nd edn. Elsevier, Oxford
20. Brix J (1993) Manoeuvring technical manual. Seehafen Verlag, Hamburg
21. Hwang W, Jakobsen B, Barr R, Ankudinov V, Fuller N, Vest L, Morris M, McGovern A, Landsburg A (2003) An exploratory study to characterize ship manoeuvring performance at slow speed. In: International conference on marine simulation and ship maneuverability (MARSIM '03), Kanazawa, Japan, vol 3
22. Yoshimura Y, Kose K, Hiraguchi T (2000) Criteria for yaw-checking and course-keeping abilities in IMO's interim standards for ship manoeuvrability. In: International conference on marine simulation and ship maneuverability (MARSIM '00). Orlando, USA, pp 389–400
23. Dand I (2003) Low-speed manoeuvring criteria: Some considerations. In: International conference on marine simulation and ship maneuverability (MARSIM '03), Kanazawa, Japan

24. Quadvlieg F, Van Coevorden P (2003) Manoeuvring criteria: More than IMO A751 require-
    ments alone! In: International conference on marine simulation and ship maneuverability
    (MARSIM '03), Kanazawa, Japan, vol 2, pp 1–8
25. International Maritime Organization (2012) SOLAS consolidated edition 2012: consolidated
    text of the international convention for the safety of life at sea, 1974, and its Protocol of 1988:
    Articles, annexes and certificates
26. Spyrou KJ (1994) A general model of ship manoeuvrability assessment based on decisions'
    analysis, and its practical application. J Soc Naval Archit Jpn 176:267–280
27. Belenky V, Falzarano J (2006) Rating-based maneuverability standards. In: SNAME annual
    meeting conference. Florida, USA, pp 227–246
28. Barr RA, Miller ER, Ankudinov V, Lee FC (1981) Technical basis for maneuvering performance
    standards. Technical Report, U.S. Department of Transportation, U.S. Coast Guard, Office of
    Merchant Marine Safety
29. International Towing Tank Conference (2002) ITTC recommended procedures and guidelines
    7.5-04-02-01: Full Scale measurements manoeuvrability full scale manoeuvring trials
30. Rijkswaterstaat (2011) Waterway guidelines 2011
31. Baudu H (2014) Ship handling, 1st edn. DOKMAR Maritime Publishers BV, Enkhuizen

Chapter 8
Impacts of Rudder Configurations on Ship Maneuvering Performance

The performance of ship rudders affects ship maneuverability, including, but not limited to, turning ability, yaw checking ability, and course keeping ability as discussed in Chap. 5. Howe ver, the existing empirical formulas for the rudder forces do not cover the impacts of the rudder profile, the rudder parameters, and the rudder interaction. Thus, the applied rudder hydrodynamics were calculated through the presented RANS methods in Chap. 4. Furthermore, an integrated maneuvering model was introduced in Chap. 6. This chapter discusses the impacts of the rudder configurations, more specifically the rudder profiles (Sect. 8.1), the spacing between twin rudders (Sect. 8.2), and the spacing among quadruple rudders (Sect. 8.3), on ship maneuverability through simulations.

Five ship types (Table 6.1) and nine rudder profiles (Fig. 2.4) are tested and compared in four chosen maneuvers from Chap. 7. Four types of maneuvering tests, namely turning, zigzag, hard turning, and lane changing are selected from the existing and proposed maneuvers. Reference seagoing and inland ships with different rudder configurations are tested, namely a single-propeller single-rudder KVLCC2 tanker (SPSR S1), a twin-propeller twin-rudder tanker (TPTR Y1) and a twin-propeller twin-rudder cargo vessel (TPTR Y2) from the Yangtze River, and a single-propeller twin-rudder ship (SPTR R1) and a twin-propeller quadruple-rudder ship (TPQR R2) from the Rhine. This chapter concludes insights into the impacts of rudder configurations on ship maneuverability. Accordingly, guidance on rudder configurations is provided in Chap. 9.

8.1 Rudder Profiles

This section presents the impacts of rudder profiles on ship maneuverability. With the proposed integrated maneuvering model in Chap. 6, the impacts of rudder profiles are illustrated by comparing the maneuvering indices of ships equipped with different

© Springer Nature Switzerland AG 2020
J. Liu, *Mathematical Modeling of Inland Vessel Maneuverability Considering Rudder Hydrodynamics*, https://doi.org/10.1007/978-3-030-47475-1_8

profiles. In addition, the spacing between the twin rudders (y_{TR}) of TPTR Y1 and TPTR Y2 (3.8 C_R and 4.0 C_R respectively) are too large to expect significant rudder interactions. Furthermore, y_{TR} of SPTR R1 and TPQR R2 is around 0.90 C_R and y_{TU} of TPQR R2 is 1.75 where the interactions are not very significant. In Sects. 8.2 and 8.3, the interactions among multiple rudders on ship maneuverability are discussed.

8.1.1 Impacts of Rudder Profiles on Turning

Figure 8.1 shows the trajectories of the reference KVLCC2 tanker (SPSR S1) with different rudder profiles in $-15°$, $15°$, $-35°$ and $35°$ turning maneuvers as an example for all the ship types. As discussed in Chap. 4, the wedge-tail series can induce a larger normal force than the IFS series and the NACA series. Therefore, the turning circles get smaller when the profiles are changed from the NACA series to the wedge-tail

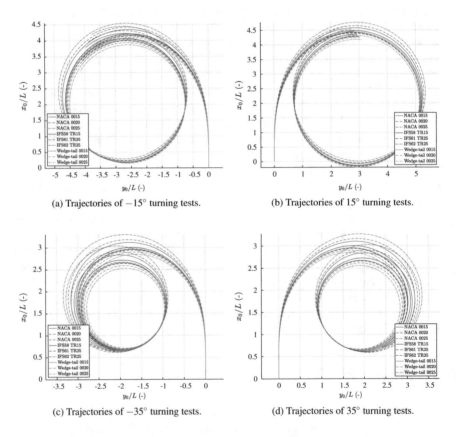

(a) Trajectories of $-15°$ turning tests.

(b) Trajectories of $15°$ turning tests.

(c) Trajectories of $-35°$ turning tests.

(d) Trajectories of $35°$ turning tests.

Fig. 8.1 Turning trajectories of a KVLCC2 tanker with different rudder profiles

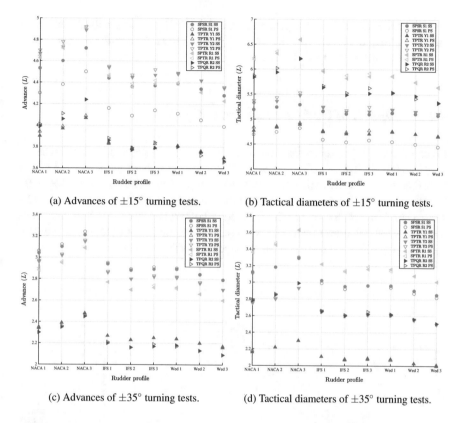

(a) Advances of ±15° turning tests.

(b) Tactical diameters of ±15° turning tests.

(c) Advances of ±35° turning tests.

(d) Tactical diameters of ±35° turning tests.

Fig. 8.2 Turning criteria of the five reference ships with different rudder profiles

series. This reduction of the turning circles is more significant for the −35° and 35° turning maneuvers than the −15° and 15° ones.

To improve the turning performance, the advance and the tactical diameter should be minimized and thus the wedge-tail series is the best option for turning. The tactical diameter and the advance of the tested KVLCC2 tanker with the best option of the compared profiles, which is the wedge-tail 0025 profile, is about 15 and 10% smaller respectively than those with the standard NACA 0018 profile.

Considering the impacts of the rudder profiles on different ships, Fig. 8.2 compares the advances and the tactical diameters of the five reference ships with nine profiles. The tendencies of the impacts of the rudder profile on ship maneuverability are similar for the tested ships. Effective rudder profiles like the wedge-tail series improve the turning ability of the tested ships. The amounts of improvement in the advances and tactical diameters from the worst case to the best one are similar, which are about 15 and 10% respectively.

In general, an increase in the number of rudders reduces the advance and the tactical diameter. The differences in the port and the starboard sides turning should be noted due to the asymmetric propeller force (SPSR S1). Additionally, the advances and the tactical diameters of $-15°$ and $15°$ turning tests are about 1.5 times larger than those of the $-35°$ and $35°$ tests.

8.1.2   Impacts of Rudder Profiles on Zigzag

Figures 8.3 and 8.4 compare the time histories of rudder angles and heading angles in $-10°/-10°$, $10°/10°$, $-20°/-20°$, and $20°/20°$ zigzag tests of the KVLCC2 tanker (SPSR S1) as a representative of the reference ships. The results show that SPSR S1 with the wedge-tail series responds faster than that with the IFS series and the NACA series. Furthermore, SPSR S1 with the wedge-tail rudders and the IFS rudders have smaller overshoot angles than ships with the NACA rudders. Furthermore, the time that takes from the first overshoot to the second overshoot is decreased significantly.

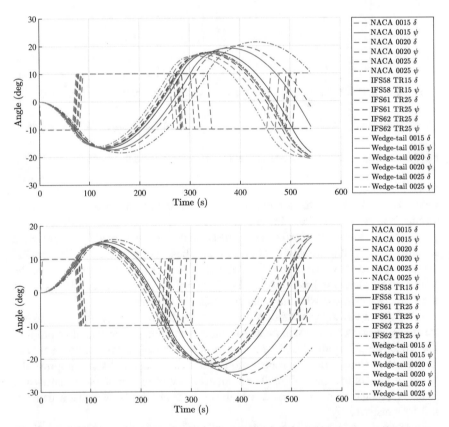

Fig. 8.3 Heading angles and rudder angles of the KVLCC2 tanker with different rudder profiles in port and starboard $10°/10°$ zigzag tests

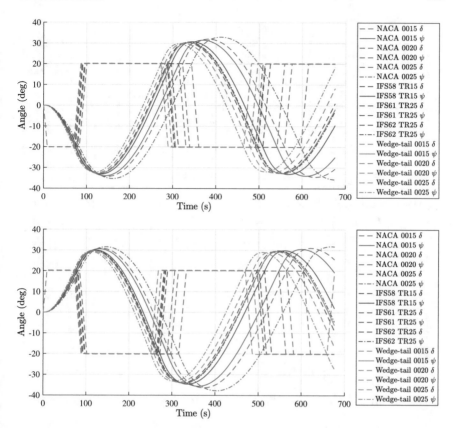

Fig. 8.4 Heading angles and rudder angles of the KVLCC2 tanker with different rudder profiles in port and starboard 20°/20° zigzag tests

Figure 8.5 shows the impacts of rudder profiles on the overshoot angles of different ships in various zigzag maneuvers. The impacts of the ship type is more significant than the rudder profile on zigzag overshoot angles which is logical because the change in the ship types (the number of rudders) increases or decreases the total rudder forces by hundreds of percent while the change in the profiles increases or decreases the total rudder forces by tens of percent.

The SPSR S1 ship shows noticeable differences with different rudder profiles in overshoot angels on port and starboard sides while the differences of the other four ships are smaller. Furthermore, the differences caused by the ship type get smaller when the applied rudder angle gets larger. The first and second overshoot angles can be reduced by about 30% when the most effective profile, wedge-tail 0025, is equipped instead of the standard NACA 0018 profile.

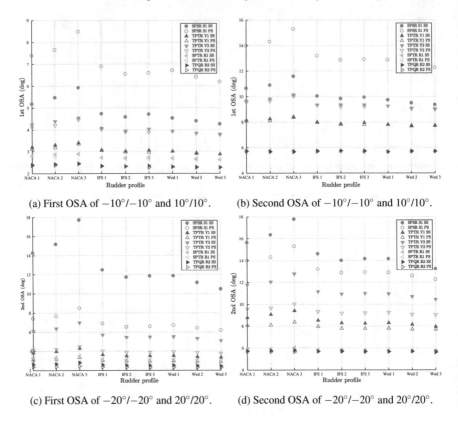

(a) First OSA of −10°/−10° and 10°/10°.   (b) Second OSA of −10°/−10° and 10°/10°.

(c) First OSA of −20°/−20° and 20°/20°.   (d) Second OSA of −20°/−20° and 20°/20°.

Fig. 8.5 Zigzag criteria of the five reference ships with different rudder profiles

8.1.3   Impacts of Rudder Profiles on Hard Turning

Figure 8.6 illustrates the impacts of rudder profiles on the hard turning trajectories of the KVLCC2 tanker. The hard turning maneuver is previously defined in Sect. 7.3.2. The end point of a trajectory is where the ship heading angle changes 90°, which is the advance position of a full turning test. The advance and the transfer increase as the rudder effectiveness decreases. For the KVLCC2 tanker, the advance and the transfer of −15° and 15° hard turning are about 30% larger than those of −35° and 35° hard turning.

Figure 8.7 presents the impacts of rudder profiles and the ship type on the transfer and the operation time in the hard turning tests. The transfer increases as the profile changes from wedge-tail to NACA and from thin to thick because the rudder forces decrease accordingly. The reference ships take more time to perform hard turning with small rudder angles (−15° and 15°) than those with large rudder angles (−35° and 35°). Furthermore, the operation time is less when more effective profiles are

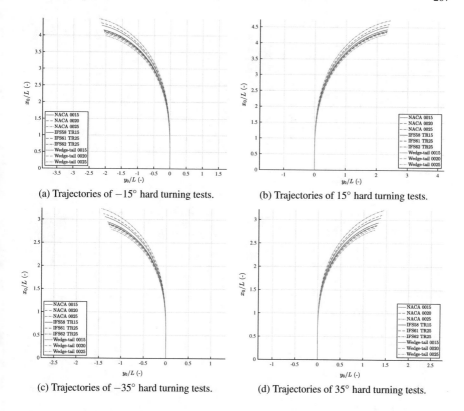

Fig. 8.6 Trajectories of the KVLCC2 tanker with different rudder profiles in hard turning tests

applied like the wedge-tail series. In general, both the transfer and the operation time can be reduced by about 10% by changing the rudder profile.

8.1.4 Impacts of Rudder Profiles on Lane Changing

Figure 8.8 shows the trajectories of the KVLCC2 tanker with different profiles in lane changing tests. The lane changing test is defined in Sect. 7.3.4. The end point of a lane changing trajectory is where the transfer is reached in a normal zigzag test. The transfer should be carefully considered when ships perform lane changing operations like overtaking in narrow waterways. The differences in the trajectories are more significant in the lane changing tests with larger rudder angles ($-20°$ and $20°$) than those with small rudder angles ($-10°$ and $10°$).

Figure 8.9 compares the transfer and the operation time of different ship types with different rudder profiles. Among the reference ships, TPQR R2 achieves the smallest transfer owing to the shortest operation time. A smaller transfer means

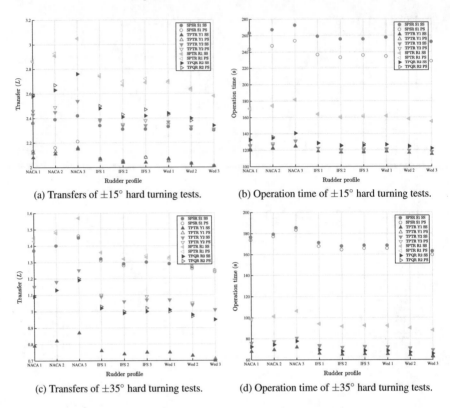

(a) Transfers of ±15° hard turning tests.

(b) Operation time of ±15° hard turning tests.

(c) Transfers of ±35° hard turning tests.

(d) Operation time of ±35° hard turning tests.

Fig. 8.7 Hard turning criteria of the five reference ships with different rudder profiles

a lower possibility of ship-bank collision in emergency lane changing operations. Furthermore, a shorter operation time presents a quicker response of the ship to the rudder order, which shows good maneuvering performance. Depending on the ship and the applied rudder angle for lane changing, the lateral deviations and the operation time can be reduced by about 30 and 20% respectively.

8.1.5   Concluding Remarks of Rudder Profiles

Section 8.1 has discussed the impacts of rudder profiles on turning, zigzag, hard turning, and lane changing maneuvers. The trajectories of the KVLCC2 tanker (SPSR S1) are taken as examples and the maneuvering criteria of other reference ships are compared. In general, reference ships with high-lift profiles like the IFS series and the wedge-tail series commonly have better maneuvering indices than the more efficient NACA series. For instance, turning indices can be reduced by about 10% by using a thin wedge-tail profile instead of a thick NACA profile.

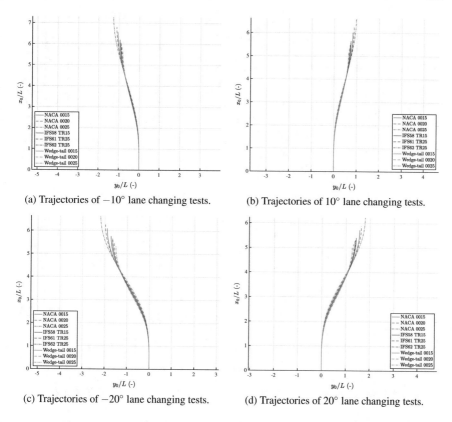

(a) Trajectories of $-10°$ lane changing tests.

(b) Trajectories of $10°$ lane changing tests.

(c) Trajectories of $-20°$ lane changing tests.

(d) Trajectories of $20°$ lane changing tests.

Fig. 8.8 Trajectories of the KVLCC2 tanker with different rudder profiles in lane changing tests

The impacts of rudder profiles on day-to-day normal operations are not so big, but in an emergency, for example, a $35°$ turn, an effective rudder profile can significantly reduce the tactical diameter, which is important for navigation safety. The impacts of the profiles on the maneuvering indices of the tested ships are similar. Additionally, ships with more propellers and rudders tend to have better maneuvering performance than those with fewer propellers and rudders. Therefore, it is recommended to use high-lift profiles and multiple-propeller multiple-rudder configurations from the perspective of good maneuverability.

8.2 Spacing Between Twin Rudders

Based on the analyzes of the spacing impacts on twin-rudder hydrodynamics in Sect. 4.2.4, this section presents the impacts of spacing between twin rudders on three reference ships, which are TPTR Y1, TPTR Y2, and SPTR R1 with the

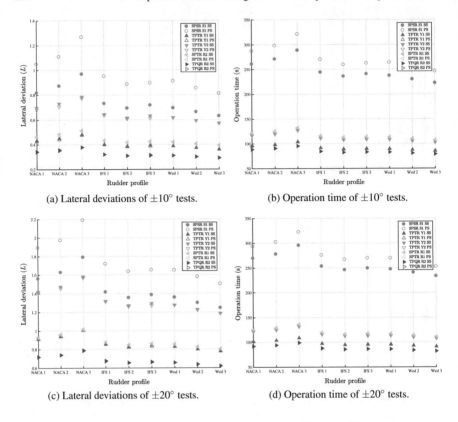

(a) Lateral deviations of ±10° tests.

(b) Operation time of ±10° tests.

(c) Lateral deviations of ±20° tests.

(d) Operation time of ±20° tests.

Fig. 8.9 Lane changing criteria of the five reference ships with different rudder profiles

NACA 0018 profile. The test range of the non-dimensional spacing between rudders $(y'_{TR} = y_{TR}/C_R)$ are 0.5 C_R to 1.0 C_R. Similar to the previous section, maneuvering simulations are performed in turning (Sect. 8.2.1), zigzag (Sect. 8.2.2), hard turning (Sect. 8.2.3), and lane changing (Sect. 8.2.4) maneuvers.

8.2.1   Impacts of Spacing Between Twin Rudders on Turning

Figure 8.10 illustrates the advances and the tactical diameters of the reference ships with different spacing between the twin rudders (y'_{TR}) in 15° and 35° turning maneuvers on port and starboard sides. As y'_{TR} increases from 0.5 to 1.0, the advances and the tactical diameters decrease about 5%. The impacts of the spacing on SPTR R1 are slightly more significant than TPTR Y1 and TPTR Y2. Additionally, the ship type does not change the tendency of the impacts of the spacing on the turning indices.

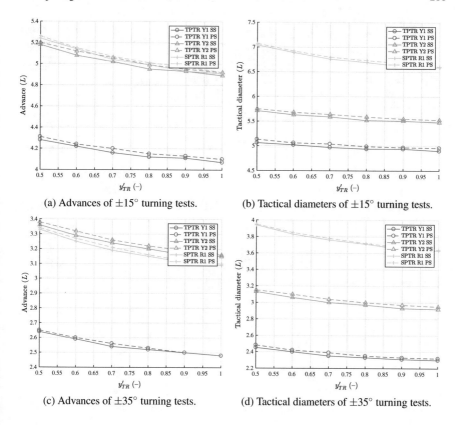

Fig. 8.10 Turning criteria of twin-rudder ships with different spacing

Basically, by changing y'_{TR}, both the advances and the tactical diameters of the tested ships in turning maneuvers can be reduced by about 5%.

8.2.2 Impacts of Spacing Between Twin Rudders on Zigzag

The first and second overshoot angles of TPTR Y1, TPTR Y2, and SPTR R1 with different spacing in 10°/10° and 20°/20° are compared in Fig. 8.11. As y'_{TR} decreases, the first and second overshoot angles of the three reference ships get smaller. Changes in overshoot angles are not very significant for TPTR Y1 and SPTR R1 (normal bow and blunt stern) but notable for TPTR Y2 (bulbous bow and slender stern). Additionally, the overshoot angles are almost the same on the starboard and port side zigzag maneuvers for the tested ships. The presented results show that both the first and the second overshoot angles can be reduced by within 5%.

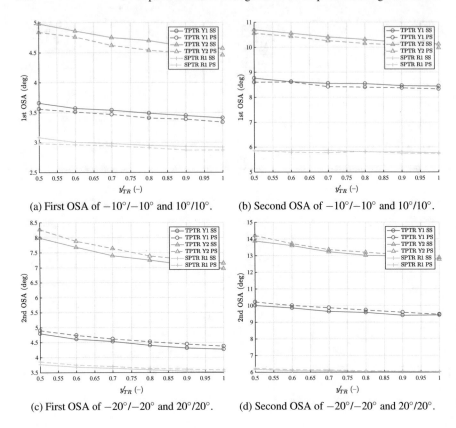

Fig. 8.11 Zigzag criteria of twin-rudder ships with different spacing

8.2.3 Impacts of Spacing Between Twin Rudders on Hard Turning

Figure 8.12 presents the transfer and the operation time of different ships with various spacing between twin rudders. As y'_{TR} increases, both the transfer and the operation time decreases. Among the three tested ships, SPTR R1 takes the longest time to change the heading angle by 90°, leading to the largest transfer. Considering the transfer and the operation time, TPTR Y1 achieves the best performance of both aspects. After all, Fig. 8.12 shows that a 5% reduction in the transfer and the operation time is possible to be achieved by changing y'_{TR}.

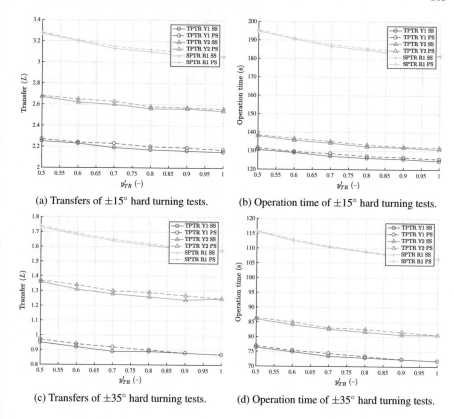

Fig. 8.12 Hard turning criteria of twin-rudder ships with different spacing

8.2.4 Impacts of Spacing Between Twin Rudders on Lane Changing

Figure 8.13 compares the lateral deviation and the operation time of three reference ships in lane changing maneuvers. The lateral deviation decreases as y'_{TR} gets larger, which means safer maneuvering in restricted waterways. Among the tested ships, TPTR Y2 has the largest lateral deviation but smaller operation time than TPTR Y1, which shows that TPTR Y2 has a better yaw checking ability and a faster response to the rudder order than TPTR Y1. After all, impacts in y'_{TR} is good for all the tested reference ships as a large spacing can reduce both the lateral deviation and the operation time. Possible reductions in the lateral deviation and the operation time by changing y'_{TR} are about 5–10%.

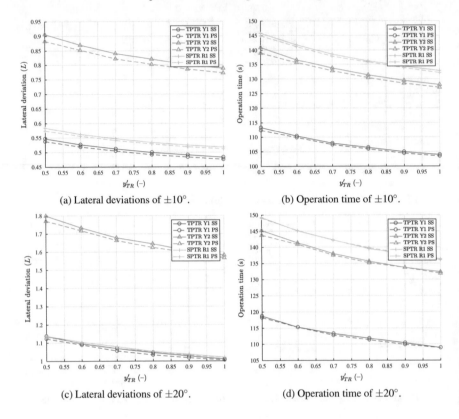

(a) Lateral deviations of $\pm10°$.

(b) Operation time of $\pm10°$.

(c) Lateral deviations of $\pm20°$.

(d) Operation time of $\pm20°$.

Fig. 8.13 Lane changing criteria of twin-rudder ships with different spacing

8.2.5 Concluding Remarks of Spacing Between Twin Rudders

Section 8.2 has presented the impacts of the spacing between twin rudders on three reference ships in four selected test maneuvers. In general, as the spacing increases in the range of $0.5\ C_R$ to $1.0\ C_R$ the reference ships react faster (shorter operation time) and, furthermore, the maneuvering performance becomes better (smaller advances, transfers, overshoot angles, and lateral deviations). These results are observed with the NACA 0018 profile. Based on the analysis of the profile impacts on ship maneuverability in Sect. 8.1, the spacing impacts with other profiles on ship maneuverability are expected to be similar.

8.3 Spacing Among Quadruple Rudders

This section focuses on the impacts of the spacing among quadruple rudders, which are the spacing between the two twin-rudder units (y_{TU}) and the spacing between the twin rudders in each twin-rudder unite (y_{TR}), on quadruple-rudder ship maneuverability. Here, the analysis is performed with a 110 m benchmark quadruple-rudder from the Rhine River (TPQR R2) as shown in Table 6.1. Similar to the previous sections, test maneuvers are turning (Sect. 8.3.1), zigzag (Sect. 8.3.2), hard turning (Sect. 8.3.3), and lane changing (Sect. 8.3.4). The test profile is NACA 0018. Additionally, y_{TU} and y_{TR} are varied in 3 steps.

8.3.1 Impacts of Spacing Among Quadruple Rudders on Turning

Figure 8.14 presents the maneuvering indices of TPQR R2 in 15° and 35° turning tests on port and starboard sides, where y_{TU} and y_{TR} are non-dimensionalized with C_R. As y_{TU} and y_{TR} increase, the rudder hydrodynamic performance gets better as previously discussed in Sect. 4.2.5. Therefore, the advances and the transfers are reduced with increases in y'_{TU} and y'_{TR}. The impacts of y'_{TU} and y'_{TR} are more significant on the 15° turning maneuvers than the 35° turning maneuvers as the operation time of the 15° turning maneuvers is larger than that of the 35° turning maneuvers. In general, the advances and the tactical diameters can be reduced about 15% by enlarging y'_{TU} and y'_{TR} from 0.5 and 0.5 to 1.5 and 1.0 respectively.

8.3.2 Impacts of Spacing Among Quadruple Rudders on Zigzag

Figure 8.15 presents the first and second overshoot angles of TPQR R2 with different y'_{TU} and y'_{TR} in −10°/−10°, 10°/10°, −20°/−20°, and 20°/20° zigzag tests. The impacts of y'_{TU} and y'_{TR} on the first overshoot angles are smaller than those on the second overshoot angles, especially for the first overshoot angles of −20°/−20° and 20°/20° which are different within 1%. Basically, as y'_{TU} and y'_{TR} become larger, overshoot angles get smaller. The possible decrease in overshoot angles caused by enlarging y'_{TU} and y'_{TR} can be 5–10%.

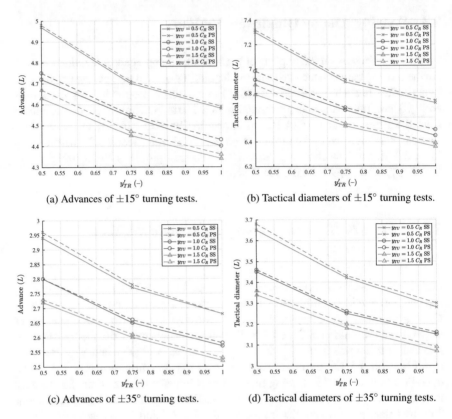

Fig. 8.14 Turning criteria of quadruple-rudder ships with different spacing

8.3.3 Impacts of Spacing Among Quadruple Rudders on Hard Turning

Figure 8.16 shows the transfer and the operation time of TPQR R2 in $\pm 15°$ and $\pm 35°$ hard turning maneuvers. As y'_{TU} and y'_{TR} get larger, both the transfer and the operation time decrease. With the same y'_{TR}, the differences of the transfer and the operation time are larger by increasing y'_{TU} from 0.5 C_R to 1.0 C_R than increasing y'_{TU} from 1.0 C_R to 1.5 C_R, which is caused by the stronger interaction effects on the rudder hydrodynamics due to smaller spacings. Additionally, the asymmetry of the port side and the starboard side maneuvering is not significant. In summary, by enlarging y'_{TU} and y'_{TR}, the transfer and the operation time of the tested quadruple-rudder ship can be reduced by about 10%.

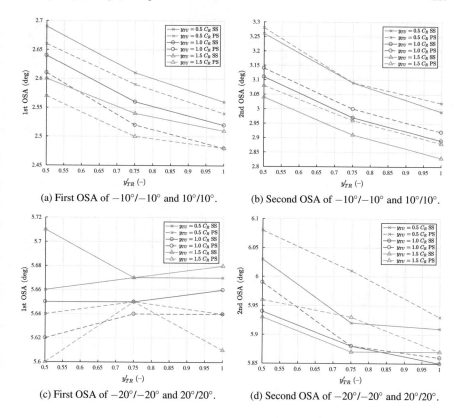

(a) First OSA of $-10°/-10°$ and $10°/10°$.   (b) Second OSA of $-10°/-10°$ and $10°/10°$.

(c) First OSA of $-20°/-20°$ and $20°/20°$.   (d) Second OSA of $-20°/-20°$ and $20°/20°$.

Fig. 8.15   Zigzag criteria of quadruple-rudder ships with different spacing

8.3.4   Impacts of Spacing Among Quadruple Rudders on Lane Changing

Figure 8.17 shows the impacts of changing y'_{TU} and y'_{TR} on the lateral deviation and the operation time of TPQR R2 in $10°$ and $20°$ lane changing on port and starboard sides. The decrease caused by increasing y'_{TU} and y'_{TR} in the lateral deviation is 10% to 15%. Accordingly, the operation time can be reduced by about 10%. The decreases in the lateral deviation and the operation time can be taken as a reduction of the risks of overtaking in narrow waterways.

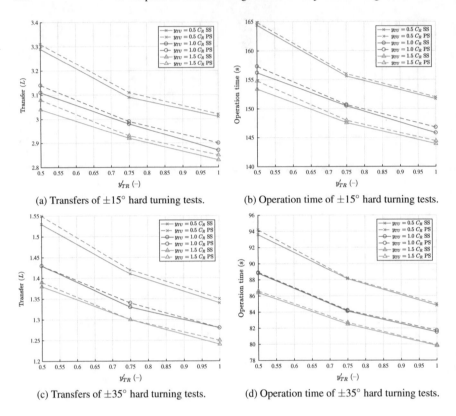

(a) Transfers of ±15° hard turning tests.

(b) Operation time of ±15° hard turning tests.

(c) Transfers of ±35° hard turning tests.

(d) Operation time of ±35° hard turning tests.

Fig. 8.16 Hard turning criteria of quadruple-rudder ships with different spacing

8.3.5   Concluding Remarks of Spacing Among Quadruple Rudders

Section 8.3 has studied the impacts of spacing among quadruple rudders on the reference ship TPQR R2 in turning, zigzag, hard turning, and lane changing maneuvers. As the non-dimensional spacing between the twin-rudder units (y'_{TU}) and the spacing between the twin rudders (y'_{TR}) of each unit increase, the maneuvering criteria like the advance, the tactical diameter, the lateral deviation, and the operation time decrease. Therefore, a better maneuvering performance is achievable by enlarging y'_{TU} and y'_{TR}.

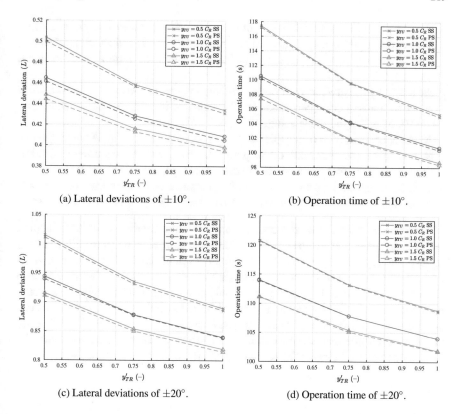

Fig. 8.17 Lane changing criteria of quadruple-rudder ships with different spacing

8.4 Summary

This chapter applied the integrated maneuvering model (Chap. 6) with the rudder hydrodynamic characteristics of different single-rudder, twin-rudder, and quadruple-rudder configurations (Chap. 4) in the classic and proposed turning, zigzag, hard turning, and lane changing maneuvers (Chap. 7). The presented results show the impacts of rudder configurations and conclude this chapter as follows:

- Various rudder profiles have different hydrodynamic characteristics, which eventually affect ship maneuverability. By changing the rudder profile, the maneuvering indices can be improved by 10–30% in different maneuvers.
- Among the tested profiles, the wedge-tail series is most effective and the NACA series is most efficient. The IFS series achieves a balance between efficiency and effectiveness.

- Basically, profiles with better hydrodynamic performance will result in a better maneuvering performance, such as smaller turning circle trajectories, advances, tactical diameters, transfers, overshoot angles, lateral deviations, and operation time.
- As the spacing between twin rudders increases, the twin-rudder reference ships get better maneuvering performance. The maneuvering indices can be reduced by 5–15% by enlarging the spacing.
- As the spacing among quadruple rudders, especially the spacing between the two twin-rudder units, increases, the ship maneuvering performance improves. Additionally, the possible reduction of the maneuvering indices is about 10%.
- With the same rudder profile, ships with more propellers and rudders tend to have better maneuvering performance than those with fewer propellers and rudders.

Ship rudder design primarily depends on the sailing conditions and the shipping purpose. In Chap. 8, the impacts of the rudder configuration are primarily analyzed on ship maneuverability, which is one of the important aspects of rudder design. In Chap. 9, the impacts of the rudder configuration on the rudder induced resistance will be briefly discussed. Other aspects should also be considered to choose a proper rudder such as the rudder torque, the rudder impacts on the propeller performance, and the rudder cavitation. Further research is suggested to improve the efficiency of the rudder while maintaining sufficient effectiveness. It should be noted that the above conclusions are achieved based on a small range of benchmark ships.

Chapter 9
Guidance on Rudder Configurations

This chapter summarizes the research of the previous chapters to provide guidance for naval architects to choose a proper rudder configuration. Section 9.1 defines the working conditions of the rudders in practice. Sections 9.2, 9.3, and 9.4 offer guidance on the rudder profile, the rudder parameters, and the rudder type of a rudder configuration respectively. Sections 9.5 and 9.6 give reference values of the rudder impacts on maneuverability and resistance. Section 9.8 draws the conclusions of this chapter.

9.1 Working Conditions

As reviewed in Chap. 2, the rudder performance depends on its working conditions, including, but not limited to, the Reynolds number and the angle of attack. Thus, operational profiles of the ship should be considered in the process of the rudder design. A high Reynolds number can be achieved in tests by either enlarging the model size or increasing the inflow speed. Presently, model tests at high-Reynolds-number or full-scale ship tests are not practical for primary studies due to the capacity of the test facilities or high expense. Another possible approach is the high-Reynolds-number CFD study, which is applied in Chap. 4.

Considering the cost of high-Reynolds-number simulations with either model tests or CFD simulations, it is recommended to carry out tests at Reynolds numbers that are as low as possible, without compromising the quality of the results. A Reynolds number of 6×10^6 is recommended, above which the rudder hydrodynamic characteristics are not significantly affected by the change in the Reynolds number. The range of applied rudder angles is $-35°$ to $35°$, which should be the main region of

© Springer Nature Switzerland AG 2020
J. Liu, *Mathematical Modeling of Inland Vessel Maneuverability Considering Rudder Hydrodynamics*, https://doi.org/10.1007/978-3-030-47475-1_9

interest in the rudder hydrodynamics. Histograms of applied rudder angles and ship speeds like Fig. 2.2 are valuable for further studies on ship maneuvering performance and fuel consumption.

9.2 Rudder Profiles

Through the previous reviews (Chap. 2) and CFD studies (Chap. 4), it is clear that different rudder profiles have different hydrodynamic characteristics, namely lift and drag coefficients, the slope of the lift curve, the stall angle, and the lift to drag ratio. These differences in the characteristics of the profile should be carefully considered in the rudder design. Furthermore, it is recommended to build up a uniform definition of the shapes of wedge-tail and fishtail profiles to make research outcomes easier to verify and expand upon existing research. In general, the choice of the rudder profile should depend on ship particulars, operational requirements of maneuvering performance, and fuel conservation.

For seagoing ships which sail long distance and commonly have tug assistance for hard maneuvering in the port area, the efficiency of the rudder may have a higher priority than the effectiveness. Thus, on the prerequisite of sufficient rudder force for course keeping and maneuvering, highly efficient profiles are suggested, such as NACA, HSVA, and IFS. For ships that mainly sail in constrained waterways like inland vessels, the effectiveness of the rudder is more crucial than efficiency. Therefore, high-lift profiles, including fishtail and wedge-tail, are proposed. Especially, when the applicable rudder area is limited due to ship draft or water depth, high-lift profiles are favorable. If the maneuverability of the ship is the key, the effectiveness can be the primary concern while the efficiency can be sacrificed.

As a general rule, thinner rudders have better hydrodynamic performance than thicker ones. The structural requirement and the hydrodynamic performance have to be considered in coincidence with the determination of the rudder thickness. With a prerequisite of the total rudder area, large span means short chord, furthermore, a large geometric aspect ratio, which is desirable for both efficiency and effectiveness. The span, the chord, and the geometric aspect ratio are commonly limited in a routine range, but the effective aspect ratio can be enlarged by adding endplates on the tip and the root of the rudder. All things considered, lift is nearly always gained at the expense of drag. A wise decision of the rudder profile should coincide with the objective of the design of the ship.

9.3 Rudder Parameters

Besides the rudder profile, the rudder parameters specify how the rudder is shaped in 3D. The rudder performance is the end product of all these parameters. First of all, the total area should be sufficiently large as it is a determinant factor of the amount of rudder force. The reference values of the total area are given in Table 2.1. When the area of a single rudder is not sufficient, multiple-rudder configurations are recommended for inland vessels because of the limited ship draft due to the enlargement of the ship dimensions. The increase in the total rudder area leads to an improvement in the ship maneuverability but makes the rudder induced resistance larger. Therefore, for ships with a large rudder area, the selection of the rudder profile becomes even more crucial. As a summary, to maximize the ship maneuverability, large total rudder area, small thickness, large span, short chord, large geometric and effective aspect ratios are recommended.

9.4 Rudder Types

Nowadays, conventional propeller-rudder systems still hold a dominant position on merchant ships. Even though active steering devices have been developing rapidly for offshore engineering, the conventional rudders can still provide good performance regarding bollard pull, response times, and cruise behavior [1]. Following consideration of the rudder type, spade rudders and semi-skeg rudders are primarily contemporary design choices. From the perspective of hydrodynamics, spade rudders are better than semi-skeg rudders for ship maneuverability and fuel consumption. Thus, it is proposed to take the spade rudder as a first choice. However, the area of a spade rudder might be limited due to the high bending moment. Semi-skeg rudders are recommended for large seagoing ships which require an extraordinary large rudder area.

9.5 Ship Maneuverability

In Chap. 8, the impacts of rudder configurations on ship maneuverability are discussed in details. Table 9.1 provides the average maneuvering indices of the reference ships (Table 6.1) as indicators, namely the advance (A_D) and the tactical diameter (T_D) that are averaged from ±15°, ±25°, ±35° turning tests, the first and second overshoot angles (ψ_{O1} and ψ_{O2}) that are averaged from ±10°, ±15°, ±20° zigzag tests, the transfer (T_r) and the operation time (t_T) that are averaged from ±15°, ±25°, ±35° hard turning tests, the lateral deviation (L_D) and the operation time (t_L) that are averaged from ±10°, ±15°, ±20° lane changing tests.

In Table 9.1, the color legends are marked based on the values of each parameter. The best available configuration among the test cases is the darkest green and the worst is the darkest red. Other colors between the darkest green to the darkest red show the transition. These color legends are presented to help naval architects to quickly compare the performance of their choice in each aspect of the maneuvering performance.

Table 9.1 Impacts of rudder configurations on the average maneuvering indices

Ship	Profile	n_R (−)	y_{TR} (C_R)	y_{TU} (C_R)	$\overline{A_D}$ (L)	$\overline{T_D}$ (L)	$\overline{\psi_{O1}}$ (deg)	$\overline{\psi_{O2}}$ (deg)	$\overline{T_r}$ (L)	$\overline{t_T}$ (L)	$\overline{L_D}$ (L)	$\overline{t_L}$ (s)
SPSR S1	NACA 0012	1	—	—	3.65	3.93	9.28	12.73	1.76	209.37	1.33	276.31
SPSR S1	NACA 0015	1	—	—	3.66	3.94	9.37	12.84	1.76	209.94	1.34	278.11
SPSR S1	NACA 0018	1	—	—	3.69	3.97	9.49	13.19	1.78	211.74	1.37	282.39
SPSR S1	NACA 0020	1	—	—	3.72	4.00	9.66	13.59	1.79	213.43	1.41	287.47
SPSR S1	NACA 0025	1	—	—	3.84	4.09	10.45	15.20	1.84	219.29	1.58	307.41
SPSR S1	IFS58 TR15	1	—	—	3.55	3.85	8.77	11.63	1.72	204.41	1.22	261.10
SPSR S1	IFS61 TR25	1	—	—	3.48	3.79	8.50	11.04	1.69	200.92	1.15	252.31
SPSR S1	IFS62 TR25	1	—	—	3.50	3.81	8.57	11.28	1.70	202.16	1.17	255.46
SPSR S1	Wedge-tail 0015	1	—	—	3.50	3.80	8.52	11.10	1.70	202.27	1.16	254.79
SPSR S1	Wedge-tail 0020	1	—	—	3.44	3.75	8.26	10.62	1.67	199.12	1.11	246.56
SPSR S1	Wedge-tail 0025	1	—	—	3.39	3.70	8.04	10.15	1.65	196.30	1.05	238.67
TPTR Y1	NACA 0015	2	1.0	—	3.03	3.30	5.59	6.37	1.36	90.90	0.67	98.62
TPTR Y1	NACA 0018	2	1.0	—	3.06	3.33	5.62	6.47	1.37	91.80	0.68	99.93
TPTR Y1	NACA 0020	2	1.0	—	3.08	3.35	5.66	6.55	1.38	92.29	0.69	101.08
TPTR Y1	NACA 0025	2	1.0	—	3.18	3.43	5.83	6.89	1.43	94.92	0.74	106.17
TPTR Y1	IFS58 TR15	2	1.0	—	2.95	3.24	5.45	6.08	1.32	88.68	0.63	94.26
TPTR Y1	IFS61 TR25	2	1.0	—	2.90	3.21	5.35	5.89	1.30	87.61	0.60	91.80
TPTR Y1	IFS62 TR25	2	1.0	—	2.92	3.22	5.38	5.98	1.31	88.02	0.61	92.62
TPTR Y1	Wedge-tail 0015	2	1.0	—	2.92	3.22	5.33	5.89	1.31	88.35	0.61	92.54
TPTR Y1	Wedge-tail 0020	2	1.0	—	2.87	3.17	5.25	5.77	1.28	86.87	0.59	90.32
TPTR Y1	Wedge-tail 0025	2	1.0	—	2.83	3.14	5.23	5.62	1.27	85.64	0.57	88.27
TPTR Y1	NACA 0018	2	0.5	—	3.37	1.52	6.10	7.49	1.52	100.26	0.84	115.61
TPTR Y1	NACA 0018	2	0.6	—	3.31	1.49	6.02	7.32	1.49	98.62	0.81	112.66
TPTR Y1	NACA 0018	2	0.7	—	3.26	1.47	5.94	7.18	1.47	97.22	0.78	110.27
TPTR Y1	NACA 0018	2	0.8	—	3.23	1.45	5.91	7.06	1.45	96.40	0.77	108.80
TPTR Y1	NACA 0018	2	0.9	—	3.20	1.44	5.86	6.95	1.44	95.58	0.75	107.32
TPTR Y1	NACA 0018	2	1.0	—	3.18	1.43	5.82	6.89	1.43	95.00	0.74	106.25
TPQR R2	NACA 0018	4	0.5	0.5	3.84	5.25	4.16	4.67	2.31	124.97	0.76	119.05
TPQR R2	NACA 0018	4	0.5	1.0	3.65	4.96	4.12	4.55	2.17	118.50	0.70	112.12
TPQR R2	NACA 0018	4	0.5	1.5	3.56	4.85	4.10	4.51	2.11	115.93	0.68	109.35
TPQR R2	NACA 0018	4	0.8	0.5	3.62	4.93	4.12	4.54	2.15	117.77	0.70	111.30
TPQR R2	NACA 0018	4	0.8	1.0	3.48	4.73	4.07	4.44	2.05	113.07	0.65	105.95
TPQR R2	NACA 0018	4	0.8	1.5	3.41	4.63	4.07	4.41	2.00	110.87	0.63	103.57
TPQR R2	NACA 0018	4	1.0	0.5	3.51	4.77	4.08	4.46	2.07	113.95	0.66	106.93
TPQR R2	NACA 0018	4	1.0	1.0	3.38	4.58	4.06	4.38	1.97	109.70	0.62	102.32
TPQR R2	NACA 0018	4	1.0	1.5	3.32	4.50	4.05	4.36	1.93	107.82	0.61	100.20

Note:

n_R is the number of rudders.

y_{TR} is the spacing between the twin rudders of twin-rudder configurations.

y_{TU} is the spacing between the two twin-rudder units of quadruple-rudder configurations.

$\overline{A_D}$ is the averaged advances of $\pm 15°$, $\pm 25°$, $\pm 35°$ turning tests.

$\overline{T_D}$ is the averaged tactical diameter of $\pm 15°$, $\pm 25°$, $\pm 35°$ turning tests.

$\overline{\psi_{O1}}$ is the averaged first overshoot angle of $\pm 10°$, $\pm 15°$, $\pm 20°$ zigzag tests.

$\overline{\psi_{O2}}$ is the averaged second overshoot angle of $\pm 10°$, $\pm 15°$, $\pm 20°$ zigzag tests.

$\overline{T_r}$ is the averaged transfer of $\pm 15°$, $\pm 25°$, $\pm 35°$ hard turning tests.

$\overline{t_T}$ is the averaged operation time of $\pm 15°$, $\pm 25°$, $\pm 35°$ hard turning tests.

$\overline{L_D}$ is the averaged lateral deviation of $\pm 15°$, $\pm 25°$, $\pm 35°$ lane changing tests.

$\overline{t_D}$ is the averaged operation time of $\pm 10°$, $\pm 15°$, $\pm 20°$ lane changing tests.

Figures 9.1, 9.2, 9.3, and 9.4 present the impacts of the rudder configurations on the averaged maneuvering criteria in four aspects which are single-rudder profiles, twin-rudder profile, twin-rudder spacing, and quadruple-rudder spacing. These figures show how the ship maneuvering performance depends on the design of the rudder configuration. Naval architects can use these figures as a reference to quickly access the performance of their initial designs.

As a summary, to achieve a good maneuvering performance, the rudder configuration should be optimized towards the more effective rudder profiles and larger spacing among rudders. Additionally, the rudder induced resistance or the power

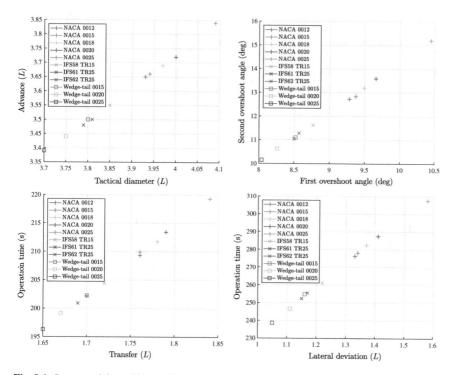

Fig. 9.1 Impacts of the rudder profiles on the average maneuvering criteria of single-rudder ships taking SPSR S1 as an example

consumed by the rudder induced resistance should be considered to have a balanced performance in both maneuvering performance and fuel consumption. Impacts of rudder configurations on ship resistance will be briefly discussed in Sect. 9.6.

9.6 Ship Resistance

In Sect. 2.2.2, Fig. 2.2 presents the probability distributions of the applied rudder angles for an inland vessel that sails from Antwerp, Belgium to Vlaardingen, the Netherlands and then from Vlaardingen, the Netherlands to Hamm, Germany. This inland vessel is almost as the same as one of the reference inland vessels, namely SPTR R1, in Table 6.1. This inland vessel equips a single ducted propeller and twin asymmetric fishtail rudders with top and bottom and plates. Based on Fig. 2.2, the probability distributions of the rudder angles in the range of $-35°$ to $35°$ are refined and shown in Fig. 9.5. This range of rudder angles is most frequently used in the two journeys, more specifically about 96 and 93% of the sailing time respectively, and thus used to consider the impacts of rudder configurations on fuel consumption.

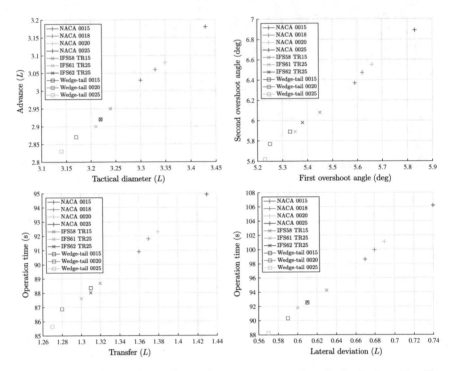

Fig. 9.2 Impacts of the rudder profiles on the average maneuvering criteria of twin-rudder ships taking TPTR Y1 as an example

Based on the probability distributions of each applied rudder angle $(P(\delta))$ in Fig. 9.5, the average applied rudder angle $(\bar{\delta})$ is calculated as the following;

$$\bar{\delta} = \sum_{i=1}^{71} P(\delta_i)|\delta_i|, \tag{9.1}$$

Accordingly, the average applied rudder angles for the routes from Antwerp, Belgium to Vlaardingen, the Netherlands $(\bar{\delta}_{AV})$ and from Vlaardingen, the Netherlands to Hamm, Germany $(\bar{\delta}_{VH})$ are 4.66° and 7.66° respectively. Afterward, these two journeys are simplified as the ship turns with constant $\bar{\delta}_{AV}$ and $\bar{\delta}_{VH}$. This simplification is made because the rudder induced forces are almost linear within the range of $-35°$ to $35°$, when the rudder is placed in the propeller slipstream as shown in Fig. 6.5.

To analyze the impacts of the rudder configurations on the fuel consumption, the first prerequisite is to ensure all the compared configurations can ensure the reference ship (SPSR R1) has the same maneuverability, more specifically the same amount of rudder induced maneuvering force. In the following study, the lift of each configuration is kept the same to calculate the average applied rudder angle $(\bar{\delta}_{AV}$

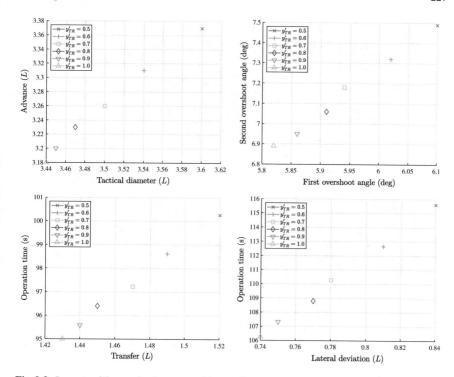

Fig. 9.3 Impacts of the spacing between rudders on the average maneuvering criteria of twin-rudder ships taking TPTR Y1 as an example

and $\bar{\delta}_{VH}$), then the rudder induced resistance is calculated. Additionally, SPTR R1 is taken as the reference ships because it has similar main dimensions to the vessel that is used to log the applied rudder angles.

Based on the hydrodynamic coefficients which are presented in Chap. 4, the average applied angles ($\bar{\delta}_{AV}$ and $\bar{\delta}_{VH}$) are calculated for each rudder configuration assuming it is fitted on SPTR R1. The power consumed by the rudder induced resistance (P), the percentage of the power consumed by the rudder induced resistance to the total power consumed by the hull and the rudder (η), and the relative difference of the power consumed by each configuration compared to the benchmark fishtail configuration ($\Delta(P)$) are calculated and compared in Table 9.2. The color legends are similar to those for Table 9.1, where the darkest green indicates the best values while the darkest red shows the worst ones.

Comparing the two routes from Antwerp, Belgium to Vlaardingen, the Netherlands and from Vlaardingen, the Netherlands to Hamm, Germany, the first one is easier for a ship to sail than the second one because small angles are more frequently applied in the first route than the second one (Fig. 9.5). Therefore, these two routes are taken as examples of which the rudder is moderately used and heavily used respectively. Accordingly, Figs. 9.6 and 9.7 show the impacts of the rudder configuration on

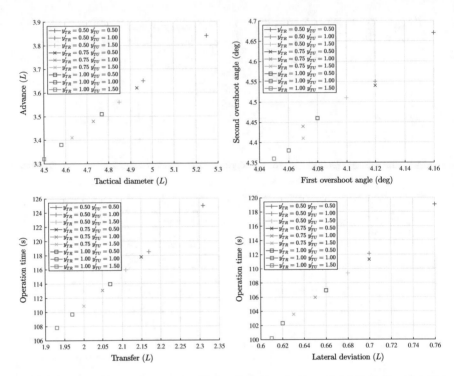

Fig. 9.4 Impacts of the spacing among rudders on the average maneuvering criteria of quad-rudder ships taking TPQR R2 as an example

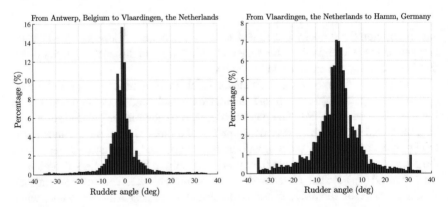

Fig. 9.5 Probability distributions of the rudder angles in the range of $-35°$ to $35°$

Table 9.2 Impacts of rudder configurations on the power consumed by the rudder induced resistance

Ship	Profile	n_R (-)	y_{TR} (C_R)	y_{TU} (C_R)	δ_{AV} (deg)	P_{AV} (kW)	η_{AV} (%)	$\Delta(P_{AV})$ (%)	δ_{VH} (deg)	P_{VH} (kW)	η_{VH} (%)	$\Delta(P_{VH})$ (%)
SPTR R1	Fishtail	2	1.00	—	4.66	16.77	13.25	0.00	7.66	25.79	19.02	0.00
SPSR R1	NACA 0012	1	—	—	8.73	10.64	8.83	-36.57	14.76	28.52	20.62	10.59
SPSR R1	NACA 0015	1	—	—	8.82	10.73	8.90	-36.02	14.91	28.79	20.77	11.63
SPSR R1	NACA 0018	1	—	—	9.05	11.05	9.14	-34.12	15.31	29.57	21.21	14.68
SPSR R1	NACA 0020	1	—	—	9.30	11.42	9.42	-31.93	15.76	30.47	21.72	18.16
SPSR R1	NACA 0025	1	—	—	10.41	12.80	10.44	-23.69	17.63	34.04	23.66	32.01
SPSR R1	IFS58 TR15	1	—	—	7.98	10.19	8.49	-39.28	13.45	26.49	19.43	2.73
SPSR R1	IFS61 TR25	1	—	—	7.51	10.13	8.45	-39.58	12.66	25.60	18.90	-0.73
SPSR R1	IFS62 TR25	1	—	—	7.63	9.97	8.33	-40.54	12.90	25.92	19.10	0.53
SPSR R1	Wedge-tail 0015	1	—	—	7.69	12.92	10.52	-22.99	12.92	28.28	20.48	9.69
SPSR R1	Wedge-tail 0020	1	—	—	7.25	11.84	9.73	-29.42	12.19	26.42	19.39	2.45
SPSR R1	Wedge-tail 0025	1	—	—	6.86	10.82	8.97	-35.52	11.52	24.67	18.35	-4.32
SPTR R1	NACA 0015	2	1.0	—	5.07	7.21	6.16	-57.00	8.53	17.72	13.89	-31.28
SPTR R1	NACA 0018	2	1.0	—	5.17	7.46	6.36	-55.53	8.70	18.17	14.20	-29.52
SPTR R1	NACA 0020	2	1.0	—	5.30	7.71	6.56	-54.03	8.93	18.70	14.55	-27.48
SPTR R1	NACA 0025	2	1.0	—	5.96	8.79	7.42	-47.57	10.01	21.10	16.12	-18.17
SPTR R1	IFS58 TR15	2	1.0	—	4.69	7.48	6.38	-55.38	7.86	17.16	13.52	-33.44
SPTR R1	IFS61 TR25	2	1.0	—	4.42	8.17	6.93	-51.28	7.41	17.34	13.64	-32.74
SPTR R1	IFS62 TR25	2	1.0	—	4.42	7.30	6.23	-56.50	7.42	16.55	13.09	-35.83
SPTR R1	Wedge-tail 0015	2	1.0	—	4.58	14.25	11.49	-15.04	7.66	23.44	17.59	-9.10
SPTR R1	Wedge-tail 0020	2	1.0	—	4.32	12.70	10.37	-24.26	7.24	21.44	16.33	-16.87
SPTR R1	Wedge-tail 0025	2	1.0	—	4.09	11.38	9.39	-32.18	6.84	19.68	15.20	-23.68
SPTR R1	NACA 0018	2	0.5	—	6.20	8.94	7.53	-46.68	10.45	21.74	16.53	-15.67
SPTR R1	NACA 0018	2	0.6	—	5.87	8.44	7.13	-49.70	9.90	20.57	15.78	-20.22
SPTR R1	NACA 0018	2	0.7	—	5.63	8.08	6.85	-51.82	9.48	19.72	15.22	-23.53
SPTR R1	NACA 0018	2	0.8	—	5.44	7.82	6.65	-53.39	9.16	19.07	14.80	-26.04
SPTR R1	NACA 0018	2	0.9	—	5.29	7.61	6.48	-54.63	8.91	18.57	14.46	-27.99
SPTR R1	NACA 0018	2	1.0	—	5.17	7.46	6.36	-55.53	8.70	18.17	14.20	-29.52
TPQR R1	NACA 0018	4	0.5	0.50	3.97	8.81	7.43	-47.45	6.68	17.11	13.48	-33.64
TPQR R1	NACA 0018	4	0.5	1.00	3.51	7.84	6.66	-53.26	5.90	15.20	12.16	-41.07
TPQR R1	NACA 0018	4	0.5	1.50	3.32	7.53	6.41	-55.13	5.59	14.52	11.68	-43.70
TPQR R1	NACA 0018	4	0.8	0.50	3.46	7.64	6.50	-54.47	5.81	14.88	11.93	-42.30
TPQR R1	NACA 0018	4	0.8	1.00	3.11	6.85	5.87	-59.18	5.23	13.40	10.87	-48.05
TPQR R1	NACA 0018	4	0.8	1.50	2.96	6.60	5.67	-60.67	4.98	12.85	10.48	-50.16
TPQR R1	NACA 0018	4	1.0	0.50	3.17	7.11	6.08	-57.61	5.33	13.79	11.16	-46.53
TPQR R1	NACA 0018	4	1.0	1.00	2.88	6.41	5.51	-61.79	4.85	12.51	10.23	-51.49
TPQR R1	NACA 0018	4	1.0	1.50	2.75	6.19	5.34	-63.10	4.64	12.04	9.88	-53.31

Note:

n_R is the number of rudders.

y_{TR} is the spacing between the twin rudders of twin-rudder configurations.

y_{TU} is the spacing between the two twin-rudder units of quadruple-rudder configurations.

AV denotes the route from Antwerp, Belgium to Vlaardingen, the Netherlands.

VH denotes the route from Vlaardingen, the Netherlands to Hamm, Germany.

δ is the average applied angles.

P is the power consumed by the rudder induced resistance.

η is the percentage of the power consumed by the rudder induced resistance to the total consumed power.

$\Delta(P)$ is the relative difference of the power consumed by each configuration compared to the benchmark configuration.

the average applied angle and the power consumed by the rudder induced resistance when the rudder is moderately and heavily used.

When the rudder configuration gets more efficient hydrodynamic characteristics, larger rudder angles have to be applied than those of configurations with more effective hydrodynamic characteristics to provide sufficient maneuvering forces. Therefore, an efficient configuration from the hydrodynamic point of view may consume more power than that of an effective configuration, more specifically the navigation efficiency is reduced. It is suggested to preselect a range of options based on the maneuverability requirements and then optimize the design to get lower power consumption. The values given in Table 9.2 can be used as indicators of how to choose a proper design while Figs. 9.6 and 9.7 show the tendency of the rudder configuration impacts on the power consumed by the rudder induced resistance. Furthermore, the tendencies of these impacts are expected to be similar to other reference ships.

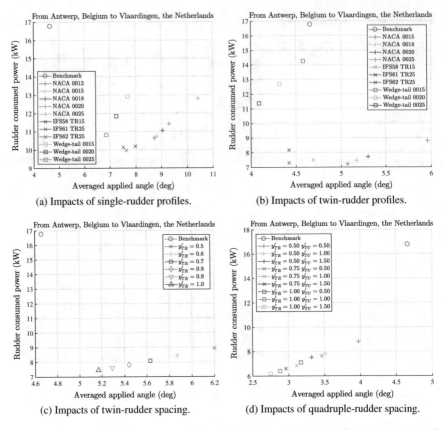

Fig. 9.6 Impacts of the rudder configuration on the average applied angle and the power consumed by the rudder induced resistance when the rudder is moderately used

9.7 Design Flow

The proposed design flow of a rudder configuration is described in Fig. 9.8. The presented hierarchy of design choices is suggested considering the effectiveness of the modification, but it is not mandatory. The design of a rudder configuration starts from the default rudder configuration, which depends on the navigation environment and the operation profile of the ship. If the ship frequently sails in bendy waterways, where the rudder has to be heavily used, the wedge-tail 0025 profile is suggested, because it is the most effective profile among the tested cases to generate the required rudder forces at relatively smaller angles. Otherwise, if the ship is designed for mostly straight moving, the NACA 0015 profile is the best choice among the studied rudder profiles because it has the lowest drag with small angles. In between, the IFS61 TR25 profile is an alternative because it achieves a balance in effectiveness and efficiency.

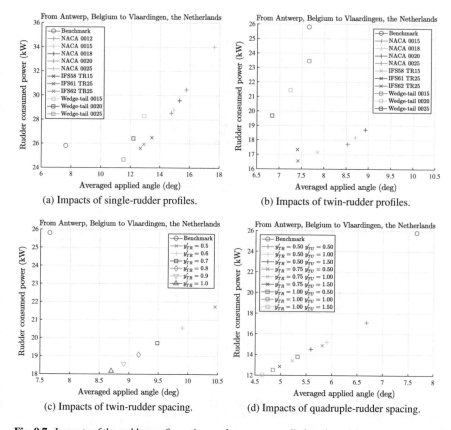

Fig. 9.7 Impacts of the rudder configuration on the average applied angle and the power consumed by the rudder induced resistance when the rudder is heavily used

The structural requirements should be checked in the selection of the rudder profile, for instance, using a thicker NACA 0018 profile instead of the thinner NACA 0015 profile. Commonly, a thicker rudder means more drag and lower hydrodynamic efficiency. A single spade type rudder with the above-mentioned default profile will be the first configuration to be considered. The default rudder area should be decided based on the reference values listed in Table 2.1 according to the type of operation. Furthermore, the rudder span should be maximized to get the largest possible aspect ratios, and then the rudder chord length is determined corresponding to the rudder area and the rudder span. In default, endplates are not applied.

After each step, the maneuvering performance of the ship should be checked against the maneuverability standards. If the default configuration cannot satisfy the requirements, the area of the single rudder is suggested to be increased first. The increase of the rudder area should be limited to 30% because the increase of the rudder area makes the aspect ratios smaller (without changing the rudder span) and the rudder heavier, which reduce the performance of the rudder and put more burden

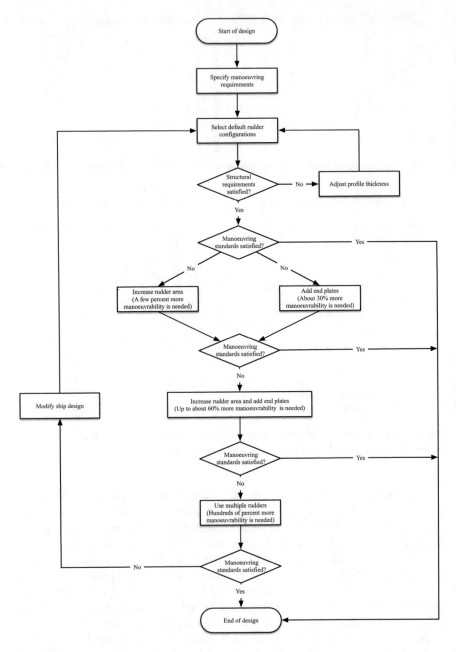

Fig. 9.8 Design flow of rudder configurations

on the steering gear. After increasing the rudder area, endplates can be applied, which may increase the lift and drag by about 30% at the same time.

If a single-rudder single-propeller system cannot satisfy the maneuvering requirements, more specifically a single rudder cannot provide sufficient maneuvering forces or an excessively large area is needed, multiple rudders should be applied, which may increase the amount of rudder forces by hundreds of percent. A multiple-rudder configuration may have a larger investment and maintenance cost than a single-rudder system. However, a ship with multiple rudders may retain part of maneuverability in case that one of the rudders is not functional, thus enhancing navigation safety. When multiple rudders are applied, the spacing between rudders should be maximized. However, the rudder should not be located out of the propeller slipstream. Otherwise, the rudder effectiveness drops sharply. If this still does not lead to sufficient maneuverability, the ship design has to be modified, namely by applying a second propeller.

9.8 Summary

The keys to improving the effectiveness of the rudder in ship maneuverability are increasing the rudder inflow velocity or the percentage of the rudder area in the propeller slipstream, enlarging the total rudder area, and improving the rudder hydrodynamic characteristics by changing the profile, the parameters, and the type. However, the improvement in the ship maneuverability in all calculated cases comes at the expense of extra rudder induced resistance. As a summary, this chapter concludes how to choose a proper rudder configuration according to the required maneuvering performance as follows:

- The working conditions should be carefully defined before choosing a configuration. The navigation conditions should be studied to estimate the probability distributions of the applied rudder angles, for instance, the probability distributions as shown in Fig. 9.5. According to the actual use of rudder angles, the configuration can be optimized to balance the maneuvering performance and fuel consumption.
- The rudder profile, the rudder parameters, and the rudder type should be considered as a whole because each choice of the rudder configuration affects its hydrodynamic characteristics, leading to a change in the rudder induced maneuvering forces and resistance.
- To improve the ship maneuvering performance, the configuration should be optimized towards larger rudder induced side force, such as using effective profiles, enlarging the total rudder area, and increasing spacing among the rudders.
- The rudder configuration should be first determined according to the requirements of maneuverability and then refined to have lower rudder induced resistance. Based on the design flow that is presented in Fig. 9.8, a proper rudder configuration can be determined accordingly considering the purpose of the ship.

- The navigation efficiency of ships with more efficient but less effective configurations from a hydrodynamic point of view may be lower than those with more effective but less efficient configurations because larger rudder angles have to be applied to provide sufficient maneuvering forces.

Reference

1. Lehmann D (2012) Station keeping with high-performance rudders. In: Dynamic positioning conference 2012. Houston, USA

Index

A

Added mass, 137
Added moment, 137
American Bureau of Shipping, 184
Angle of attack, 14, 16

B

Body-fixed coordinate sytem, 135
Boundary conditions, 52
Bureau Veritas (BV), 185

C

Captive model tests, 114
Cavitation, 36
Central Commission for the Navigation of
    the Rhine (CCNR), 184
Computational Fluid Dynamics (CFD), 14
Clam shell angles, 32
Conservation of energy, 47
Conservation of mass, 46
Conservation of momentum, 46
Coordinate systems, 109
Course keeping ability, 183, 187
Critical angle of attack, 14

D

Degrees of freedom, 108
Density-based solver, 54
Discretization schemes, 54
Drag coefficient, 15
Drag force, 14, 15

E

Earth-fixed coordinate system, 135
Effective rudder angle, 16
Entry, 47
European Commission (EC), 185
Evasive action test, 188
External design factors, 100, 106

F

Fairing tip, 22
Finite difference method, 54
Finite element method, 54
Finite volume method, 54
Fishtail profiles, 21
Flat-plate profiles, 20
Flow straightening coefficient, 139
Form drag, 15
Free-running tests, 114
Friction drag, 15
Fully-balanced rudders, 27

H

Horn rudders, 27
HSVA profiles, 20
Hull forms, 107

I

IFS profiles, 21
IMO standards, 182
Initial turning ability, 183
Inland navigation, 5
Inland vessel, 1
Inland waterway transport, 2

© Springer Nature Switzerland AG 2020
J. Liu, *Mathematical Modeling of Inland Vessel Maneuverability Considering Rudder Hydrodynamics*, https://doi.org/10.1007/978-3-030-47475-1

Printed in the United States
by Baker & Taylor Publisher Services